Gene Editing, Law, and the Environment

Life Beyond the Human

Technologies like CRISPR and gene drives are ushering in a new era of genetic engineering, wherein the technical means to modify DNA are cheaper, faster, more accurate, more widely accessible, and with more far-reaching effects than ever before. These cutting-edge technologies raise legal, ethical, cultural, and ecological questions that are so broad and consequential for both human and other-than-human life that they can be difficult to grasp. What is clear, however, is that the power to directly alter not just a singular form of life but also the genetics of entire species and thus the composition of ecosystems is currently both inadequately regulated and undertheorized. In *Gene Editing, Law, and the Environment*, distinguished scholars from law, the life sciences, philosophy, environmental studies, science and technology studies, animal health, and religious studies examine what is at stake with these new biotechnologies for life and law, both human and beyond.

Irus Braverman is Professor of Law and Adjunct Professor of Geography at the University at Buffalo, The State University of New York. She is author of *Planted Flags: Trees, Land, and Law in Israel Palestine* (2009), *Zooland: The Institution of Captivity* (2012), and *Wild Life: The Institution of Nature* (2015), and co-editor of *The Expanding Spaces of Law: A Timely Legal Geography* (2014) and *Animals, Biopolitics, Law: Lively Legalities* (2016).

Law, Science and Society series

General editors

John Paterson
University of Aberdeen, UK

Julian Webb
University of Melbourne, Australia

Law's role has often been understood as one of implementing political decisions concerning the relationship between science and society. Increasingly, however, as our understanding of the complex dynamic between law, science and society deepens, this instrumental characterisation is seen to be inadequate, but as yet we have only a limited conception of what might take its place. If progress is to be made in our legal and scientific understanding of the problems society faces, then there needs to be space for innovative and radical thinking about law and science. *Law, Science and Society* is intended to provide that space.

The overarching aim of the series is to support the publication of new and groundbreaking empirical or theoretical contributions that will advance under-standing between the disciplines of law, and the social, pure and applied sciences. General topics relevant to the series include studies of:

* law and the international trade in science and technology;
* risk and the regulation of science and technology;
* law, science and the environment;
* the reception of scientific discourses by law and the legal process;
* law, chaos and complexity;
* law and the brain.

Titles in this series:

Knowledge, Technology and Law
Edited by Emilie Cloatre and Martyn Pickersgill

Law and the Management of Disasters
The Challenge of Resilience
Edited by Alexia Herwig and Marta Simoncini

Gene Editing, Law, and the Environment
Life Beyond the Human
Edited by Irus Braverman

Gene Editing, Law, and the Environment

Life Beyond the Human

Edited by Irus Braverman

Routledge
Taylor & Francis Group

LONDON AND NEW YORK

First published 2018
by Routledge

2 Park Square, Milton Park, Abingdon, Oxfordshire OX14 4RN
52 Vanderbilt Avenue, New York, NY 10017

a Glasshouse book

Routledge is an imprint of the Taylor & Francis Group, an informa business

First issued in paperback 2018

British Library Cataloguing-in-Publication Data
A catalogue record for this book is available from the British Library

Library of Congress Cataloging-in-Publication Data
Names: Braverman, Irus, 1970– editor.
Title: Gene editing, law, and the environment : life beyond the human /
    edited by Irus Braverman.
Description: New York, NY : Routledge, 2017. | Series: Law, science
    and society | Includes bibliographical references and index.
Identifiers: LCCN 2017007958 | ISBN 9781138051126 (hbk) |
    ISBN 9781315168418 (ebk)
Subjects: LCSH: Human genetics—Law and legislation. | Genetic
    recombination. | Genetic regulation. | Mutation (Biology)
Classification: LCC K3611.G46 G425 2017 | DDC 344.04/196—dc23
LC record available at https://lccn.loc.gov/2017007958

ISBN: 978-1-138-05112-6 (hbk)
ISBN: 978-0-367-13846-2 (pbk)

Typeset in Baskerville
by Apex CoVantage, LLC

For my father, *Dr. Dan Braverman*

Contents

Acknowledgments

This collection is the end product of a very unusual workshop and public lecture held at the University at Buffalo School of Law, the State University of New York, in October 2016. There was an unmistakable buzz in the room: scholars from disciplines far and wide discovered that while they may share similar political concerns and an honest desire to communicate, they needed to negotiate a common language through which to conduct these communications. Jargon needed to be minimized. Respect was key. It also may have helped that these scholars are all prominent thinkers in their respective fields, with many high achievements to their credit.

As is often the case, at this workshop, too, the most interesting observations emerged in the less formal settings. Indeed: during the first lunch break, the Science and Technology Studies (STS) people congregated in the right-hand side of the room (I called it the "STS circle"), while the animal people took their lunch to the other corner. The governance people had their own mini-discussion going, and the scientists, I noticed with a certain panic, were nowhere to be seen. I found myself feverishly moving in between the groups. Perhaps unsurprisingly, by the second lunch these boundaries had diminished, and by the end of the workshop they were almost nonexistent, and other affiliations emerged in their place. This interdisciplinarity, and the surprising relationships that have formed through it, are an important aspect of this volume.

A few words of gratitude and appreciation are in order. I will start at the beginning: my early conversations with Ben Hurlbut from Arizona State University helped to strike the tone and gather the relevant scholars for the workshop. Sheila Jasanoff of Harvard University's Kennedy School of Government read and commented on a couple of early drafts that articulated the questions framing the workshop's discussion. Sheila had also generously agreed to co-deliver the public Mitchell Lecture that was held in conjunction with the workshop, which attracted a large audience from across the university and beyond. Jeantine Lunshof of George Church's lab at Harvard Medical School was always delightfully thoughtful and supportive, and Gaymon Bennett from Arizona State University helped me through the final stretch of the editing process. Lengthy conversations with Hank Greely from Stanford University, Jake Sherkow from New York Law School, and Jim Collins from Arizona State University were instrumental in shaping the regulatory and ecological orientation of this volume.

In my own explorations of gene drives, I was fortunate to have met several unique individuals who were wonderful and patient teachers to me: Flaminia Catteruccia and Andie Smidler of Harvard University's Chan School of Public Health, and Kenneth Oye of MIT's Political Science Department. I am especially indebted to Kevin Esvelt of MIT's Media Lab, who has dedicated more hours than I can count on two hands to explain the science of gene drives to me and who so generously shared his provocative vision both at the workshop and as a co-speaker at the Mitchell Lecture. Finally, I would like to thank each and every one of this volume's contributors for their openness, honesty, and creativity and for their courageous work at the cutting edge of science, law, and the environment.

I am also thankful to my colleagues at the University at Buffalo, and to the many staff and students who have contributed to both the workshop and the Mitchell Lecture, and who have made them possible on so many levels. These include: Josephine Anstey, Guyora Binder, James Bono, Marlene Cook, Rebecca Donoghue, Ilene Fleischmann, Richard Gronostajski, Marc Halfon, Debra Kolodczak, Gerald Koudelka, Katie Little, and Paul Vanouse. My research assistant John DiMaio in particular provided invaluable help throughout the long process from organizing the workshop to turning the papers into book chapters. I am also very grateful to UB's School of Law and to its Genome, Environment and Microbiome (GEM) Community of Excellence for their financial support of the workshop and public lecture that have been foundational for this volume. Additionally, I would like to extend my gratitude to Errol Meidinger of the Baldy Center for Law & Social Policy, and to Jennifer Surtees, co-director of GEM and associate professor at the department of biochemistry, for their help in putting the workshop together. I would also like to thank artist and designer Michael Morgenstern for generously agreeing to let me use his wonderful illustrations as entries for each chapter. Michael has illustrated for many of the leading magazine and book publishers and was selected as one of the top 200 illustrators worldwide by Luerzer's Archive.

I will end by thanking my family for their incredible patience and humor in the face of my recent obsession with CRISPR and genetics. To my older daughter, Ariel, for letting me try out some basic genetic teaching skills on her class, and to my younger daughter, Tamar, who has been asking me some tough questions about life that have motivated me to study on. My life partner, Gregor Harvey, has held everything together for me, as he always does. I would like to dedicate this volume to my father, Dr. Dan Braverman, who has devoted his life to understanding and healing the human body and who has instilled in me a deep love and respect for biology. Thank you Aba.

Notes on Contributors

Lori Andrews is a graduate of Yale College and Yale Law School, and is distinguished professor at IIT Chicago-Kent College of Law. She wrote 11 non-fiction books, more than 150 articles, and three mystery books with a female geneticist protagonist. Andrews chaired the federal ethics advisory committee to the Human Genome Project.

Gaymon Bennett is professor of religion, science, and technology at Arizona State University. He is author of *Technicians of Human Dignity: Body, Souls, and the Making of Intrinsic Worth* (Fordham University Press, 2015) and coauthor of *Designing Human Practices: An Experiment in Synthetic Biology* (with Paul Rabinow) (The University of Chicago Press, 2012).

Irus Braverman is professor of law at the University at Buffalo, The State University of New York. She is author and editor of eight books, including *Wild Life: The Institution of Nature* (Stanford University Press, 2015) and *Coral Whisperers: Scientists on the Brink* (The University of Chicago Press, forthcoming).

Kevin M. Esvelt is assistant professor at Massachusetts Institute of Technology's Media Lab. He received his Ph.D. in biochemistry from Harvard University. Working at the Wyss Institute and Harvard Medical School, Esvelt developed the CRISPR gene drive in collaboration with George Church's lab. He currently leads the Sculpting Evolution group.

Stephen Hilgartner is professor at the Department of Science & Technology Studies, Cornell University. He studies the social dimensions and politics of contemporary and emerging science and technology, especially in the life sciences. His most recent book is *Reordering Life: Knowledge and Control in the Genomics Revolution* (MIT Press, 2017).

J. Benjamin Hurlbut is assistant professor at Arizona State University's School of Life Sciences. His research lies at the intersection of science and technology studies, bioethics, and political theory. He is author of *Experiments in Democracy: Human Embryo Research and the Politics of Bioethics* (Columbia University Press, 2017).

Todd Kuiken is a senior research scholar with the Genetic Engineering and Society Center, North Carolina State University. He was principal investigator at Woodrow Wilson Center's Synthetic Biology Project, worked at the National Wildlife Federation and Island Conservation, and is currently a member of the UN Convention on Biological Diversity Ad-Hoc Technical Expert Group on Synthetic Biology.

Stuart A. Newman is a biologist at New York Medical College, Valhalla, New York. He was an early contributor to the field of systems biology and identified the mechanism of vertebrate limb development. He was a founding member of the Council for Responsible Genetics and is editor of the journal *Biological Theory*.

Ronald Sandler is professor of philosophy and chair of the Department of Philosophy and Religion at Northeastern University. Among many other books, he is author of *Character and Environment* (Columbia University Press, 2009), *The Ethics of Species* (Cambridge University Press, 2012), and *Environmental Ethics: Theory in Practice* (Oxford University Press, 2018).

Alexander J. Travis is associate professor of reproductive biology at Cornell University's College of Veterinary Medicine, as well as associate dean for International Programs and director of the Master of Public Health program. In 2015, Travis's lab announced the birth of the world's first litter of puppies by *in vitro* fertilization.

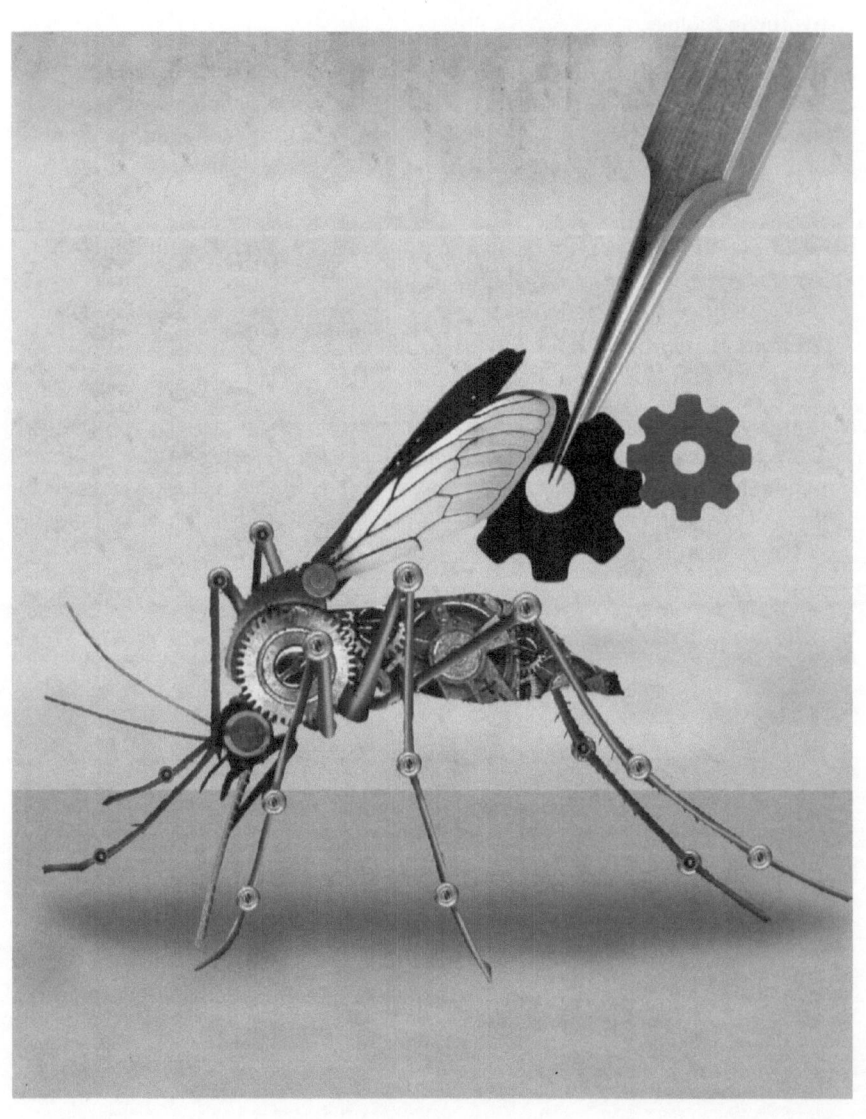

Introduction

Editing the Environment
Emerging Issues in Genetics and the Law

Irus Braverman

> [S]cience and technology govern lives as surely as law does, empowering some forms of life and making them natural while others, by comparison, come to seem deficient or unnatural . . . As in moments of lawmaking or constitutional change, the emergence of a far-reaching technology like CRISPR is a time when society takes stock of alternative imaginable futures and decides which ones are worth pursuing and which ones should be regulated, or even prevented.
>
> Jasanoff, Hurlbut, and Saha 2015, 26–27

Technologies like CRISPR—a method for editing a cell's genome—are ushering in a new era of genetic engineering, wherein the technical means to modify DNA are cheaper, faster, more accurate, and more widely accessible than ever before. Developed in 2012, CRISPR is emerging as a powerful new genome engineering technology and as a locus of international concern over the ethical and legal norms that will guide its application in the biosciences. Developed in 2014, an even newer technological application—the gene drive—uses CRISPR to alter not just individual organisms, but also the genetics of entire populations. These cutting-edge gene editing technologies raise ethical, legal, and ecological questions that are so broad and consequential for both human and more-than-human life that they can be difficult to grasp. What is clear, however, is that the power to directly alter, not just a singular form of life, but also the genetics of entire species and thus the composition of ecosystems, is currently both under-regulated and under-theorized. CRISPR and its application in gene drives serve as the central impetus for this volume, as these technologies raise myriad issues at the nexus of law, science, and the environment.

Gene Editing, Law, and the Environment examines the important relationship between genetics and the law and how this relationship impacts and is impacted by the environment. Considered from multiple perspectives, the volume's triangular focus on genetics, law, and more-than-human life allows for a nuanced exposition of how diverse disciplines are currently approaching novel technologies of gene editing: the central concepts used, the main issues flagged, and the major debates kindled. In addition to this introduction, the volume consists of nine chapters and a short afterword by distinguished scholars from an array of scholarly and professional orientations, including law, the life sciences, philosophy, environmental studies, science and technology studies, animal health, and religious studies. The contributors—who participated in a two-day workshop held in Buffalo, New York in October 2016—provide a rich tapestry of approaches that together examine what is at stake with these new biotechnologies for life and law.

The discussions here extend beyond the applications of gene editing technologies on humans to consider their effects on other life forms. Potentially spreading through populations and species and greatly impacting entire ecosystems, gene editing technologies present ethical problems, political questions, and regulatory challenges that have thus far received far too little attention in the academic literature. While the few conversations that have emerged typically treat the applications of gene editing to humans and nonhumans as separate matters, this collection draws attention to how these new technologies, and CRISPR and gene drives in particular, would apply to nonhumans and ecosystems, alongside, and in conversation with, their potential applications to humans. By exploring the range of possible applications of gene editing—not only biomedical but also agricultural, ecological, and cultural—this volume's contributors draw attention to crossovers and disjunctions between approaches to the human and to the nonhuman.

In addition to shedding new light on the human-nonhuman divide, novel gene editing technologies like CRISPR and gene drives also challenge other traditional bifurcations, such as those between nature and culture, law and science, public and private, lab science and field science, and synthetic and conservation biology. The cutting-edge nature of this topic and the interdisciplinarity of the contributions make this a one-of-a-kind collection. Indeed, in his afterword to this volume, Science and Technology Studies scholar Stephen Hilgartner draws on the term "constitutional moment" (Jasanoff 2003)—a time when the basic rules and infrastructures that establish order are either founded or fundamentally changed—to highlight what he sees as the "constitutional conversation" that takes place in this volume. And "although mere conversations lack decision-making capacity and a formal legal standing," Hilgartner points out, "they are nevertheless spaces where frameworks of governance are configured and advanced. Moreover, in some cases, such conversations make strides toward creating the shared ontologies that underwrite the frameworks that, in turn, constitute new regimes of governance" (see Afterword, page 188).

In what follows, I will briefly introduce CRISPR ("CRISPR than Life") and gene drives ("Gene Drives and the Environment"), after which I will discuss the twists and turns of these technologies' legalities ("CRISPR Legalities") and how they can be understood from the standpoint of the science–law relationship ("Law and Science: An Open-Ended Conclusion"). Throughout, I will be weaving together the myriad questions and discussions raised by the volume's diverse contributions.

CRISPR than Life

Standing for "clustered regularly interspaced short palindromic repeats," CRISPR is a naturally occurring system by which prokaryotes such as bacteria defend themselves against viruses. CRISPR systems have been found in approximately 40 percent of sequenced bacterial genomes and 90 percent of sequenced archaea (Université Paris 2017). In the last several years, scientists have modified several versions of the natural CRISPR system to edit the genomes of nucleated organisms such as animals and plants. CRISPR utilizes specific RNA molecules to direct its enzyme component, Cas, to precisely snip out a piece of DNA at almost any desired point within the genome. The cell's standard response is to repair the DNA at the break point by directly sealing up the broken ends, often deleting or inserting a few bases (i.e., adenine, cytosine, guanine, or thymine) in the process, which can disrupt the function of the gene. Alternatively, scientists can use a synthetic DNA sequence as a template for repair, thereby exactly incorporating a desired sequence into the genome at a particular location. "All CRISPR does is cut the DNA," one geneticist explained. "Everything else is the cell repair system, and that's what we're hitching on to" (Cohen 2016).

CRISPR has thus enabled what some scientists depict as an effortless editing of any living organism (Kahn 2015). The first CRISPR enzyme used to successfully edit mammalian genes was Cas9 (Brown 2017). In 2015, the Broad Institute announced a second CRISPR system, CRISPR-Cpf1, and in 2016, researchers from several universities announced the discovery of C2c2, a CRISPR enzyme that targets RNA rather than DNA. Most recently, in December 2016 researchers at Berkeley announced the discovery of two new CRISPR-Cas systems: CRISPR-CasX and CRISPR-CasY (ibid.). New and yet newer applications that enable scientists to further edit the genome are appearing almost daily.

Observing the extraordinary versatility and rapid uptake of CRISPR, the prominent journal *Science* has declared a "CRISPR Revolution." According to many, CRISPR's revolution not only pertains to the biological foundations of life, but also to social, moral, legal, and political life. As *Science* declares: "for better or worse, we all now live in CRISPR's world" (Travis 2015). Quoted in a recent issue of *The New Yorker*, bioethicist Henry Greely of Stanford University compares CRISPR to the Model T Ford: far from the first automobile, but the one whose simplicity of production, dependability, and affordability transformed society (Specter 2015). "Any molecular biology lab that wants to do CRISPR can," says Harvard University's George Church (Travis 2015). "Any idiot can do it," MIT biologist Rudolf Jaenisch, who designed the first transgenic mouse in 1974 and pioneered the engineering of CRISPR mice, qualifies (Cohen 2016).

The biomedical applications of CRISPR are proliferating at a dizzying pace. Clinical researchers are already applying CRISPR in an attempt to create tissue-based treatments for cancer and other diseases, and proposals for growing human organs in animals for human transplantation are underway (Reardon 2015; Wu et al. 2017). A 2015 international summit on CRISPR saw many discussions of its promise for repairing genetic defects in human embryos, if society dares cross what many regard as an ethical threshold: altering the human germline (Travis 2015).

Researchers have also harnessed CRISPR to design a growing menagerie of genetically engineered animals and plants: super muscular beagles (Regalado 2015b), pigs resistant to certain viruses (Whitworth et al. 2016), longer lasting tomatoes (Regalado 2015a), and wheat that can fend off a widespread fungus (Wang et al. 2014). Allergen-free peanuts, biofuel-enhanced poplar trees, and human-animal chimeras are all on the drawing board (Travis 2015; Wu et al. 2017). Unlike earlier techniques for genetically modifying organisms, CRISPR can do its work without leaving any "foreign" DNA behind. This characteristic poses a challenge for existing regulations, which are often based on the presence of such foreign DNA.

Gene Drives and the Environment

Combined with CRISPR, an even newer biotechnological application utilizes another natural phenomenon, known as a "gene drive," to rapidly alter the genetic makeup of a species so to ensure that a trait is always inherited (Esvelt et al. 2014; Specter 2017). Gene drives work by overriding the Mendelian rules of inheritance. Normally, the progeny of any sexually reproductive organism receives half its genome from each parent; but the new technological application ensures that the new gene will copy itself in every successive generation (Esvelt et al. 2014; Specter 2017; see also Esvelt, Chapter 1, this volume). The *New Yorker* explains, accordingly, that:

> a mutation that blocked the parasite responsible for malaria, for instance, could be engineered into a mosquito and passed down every time the mosquito reproduced. Each future generation would have more offspring with the trait until, at some point, the entire species would have it.
>
> (Specter 2017)

The same article also quotes Kevin M. Esvelt of MIT, who was the first to describe, in 2014, how CRISPR could combine with a gene drive to alter the genetic composition of entire species in the wild: "There has never been a more powerful biological tool, or one with more potential to both improve the world and endanger it" (ibid.).

Indeed, although the CRISPR-gene drive combination is only effective under certain conditions (for example, it requires sexual reproduction and short generation spans), this technology could grant humans the power to intentionally and rapidly engineer not only individuals but also populations and species, thereby impacting entire ecological systems. The vast array of possible environmental applications of gene drives includes preventing mosquito and tick-borne diseases such as malaria, Zika, dengue, yellow fever, and Lyme disease, eliminating invasive species that impact humans and nonhumans, increasing the genetic diversity of shrinking populations, improving agricultural productivity in a world with a growing number of humans, and facilitating adaptation to climate change.

Alongside the huge promise of this technology, debates are currently unfolding over the benefits and ecological risks of using gene drives for conservation. In his contribution to this volume, entitled *"Gene Drives and Species Conservation: An Ethical Analysis"* (Chapter 2), environmental philosopher Ronald Sandler argues that gene drives offer a new type of conservation, which he refers to as the "genetic turn." Instead of focusing on the conditions in which species are found, Sandler asserts, CRISPR-enabled interventions newly focus on the genetic constitution of the organisms themselves. Rather than mitigating the destructive effects of human actions, the biology of the species thus becomes the problem that humans can, and must, fix by engineering organisms to be more adaptable to the changing environment. In Sandler's words:

> This [approach] allows humans to adapt [wild organisms] to us and to our anthropogenic world, rather than requiring us to adapt our lifestyles and production systems to accommodate them. It is evolution by artificial selection among engineered variations: a full embrace of the Anthropocene.
>
> (Chapter 2, page 49)

Traditionally, conservation biologists have not only insisted on minimizing human interventions, but have also taken cues from historical ecosystems to guide appropriate conservation management strategies. As Sandler sees it, the genetic turn poses a fundamental dilemma for the project of conservation biology: if we are no longer concerned with conserving the natural world but are more invested in designing it, what is the meaning of conservation in the first place? Echoing this sentiment, in September 2016, a group of conservation leaders issued an open letter calling for a moratorium on research and field trials in gene drives until the International Union for Conservation of Nature—the oldest and largest conservation organization in the world—develops clear guidance on this topic (Friends of the Earth 2016).

Choosing a different entry point to the topic of gene drives and conservation than that of Sandler, environmental scientist Todd Kuiken's chapter, *"Vigilante Environmentalism: Are Gene Drives Changing How We Value and Govern Ecosystems"?* (Chapter 5), asks whether the increasing accessibility of genomic tools, combined with bottom-up mechanisms like crowdfunding, effectively instantiates a new form of environmentalism, which he refers to as "vigilante environmentalism." Vigilante environmentalism—here, gene drive projects advanced by self-appointed environmentalists-cum-biotechnologists—is a double-edged sword, Kuiken is quick to clarify. On the one hand, this process has already and will further enable the "democratization" of science and the emergence of a vibrant community of do-it-yourself biologists. However, the greatly enhanced accessibility that has developed with the advent of new gene editing technologies has purportedly also introduced increased risks and dangers into these processes. Kuiken explains that such perceived risks and dangers have set the stage for the recent massive involvement of the United States military in gene editing. What, he wonders, does this new "militant environmentalism" mean for conservation?

In a dramatic response to the potential perils of the new gene editing era, the United States President's Council of Advisors on Science and Technology (PCAST) released a letter to former President Obama. The letter, dated November 15, 2016, explains that "advanced biotechnology offers the promise of transforming the way the world treats disease, but it also has the potential for destructive use by both states and technically competent individuals with access to modern laboratory facilities" (U.S. White House 2016, 1). "It is possible that a well-planned, well-executed attack might go unnoticed for days or weeks," the security experts continue. "The ability of the United States to escape serious consequences will depend on effective detection (biosurveillance), response (such as medical countermeasures), and recovery capabilities" (ibid., 4). The challenges are considerable, the experts conclude, calling for an urgent biodefense strategy that would protect the nation from the dangers posed by the new technologies if coopted by terrorists.

Partly in response to such threats of bioterrorism, over the last ten years, the Pentagon's Defense Advanced Research Projects Agency (DARPA) has increased its involvement in synthetic biology research and is currently one of the largest public funders of this research in the United States. Inspired by the recent advances in the field of gene drives, in September 2016 DARPA announced the Safe Genes program, which aims to develop responses for accidental or malicious "genetic spills" into the environment (DARPA 2016; see also Chapter 5, this volume).

Despite these and other concerns over gene drives, only a couple of months earlier, the National Academies of Sciences, Engineering, and Medicine released a report that essentially grants a green light to highly controlled field trials with this technology (2016). Whether or not gene drives will be defined as genetically modified agricultural products, and in which contexts, will determine whether the Cartagena Protocol of 2000 on Biosafety and the Nagoya–Kuala Lumpur Supplementary Protocol of 2010—both attached to the United Nations Convention on Biological Diversity (CBD)—will apply to this technology. These two major international protocols address genetically modified organisms and pertain only to transboundary actions; they do not apply to the use or transit of genetically modified organisms within nation states.

Such discussions highlight both the necessity of, and the problems with, existing international government apparatuses in contending with genetic technologies that are not easily bound by geographic and jurisprudential distinctions. The National Academies of Sciences took a clear stance on this issue when it recommended the CBD as a platform for regulating gene drives on an international scale. Signed by nearly 200 countries—with the notable exception of the United States—the CBD was drafted to meet a series of biodiversity goals, but also includes provisions on the movement of genetically modified organisms across borders. Many saw the December 2016 meeting of the CBD in Cancún as an opportunity to bring their concerns about genetic editing to the negotiating table. But while the idea of a moratorium found support among some countries, the final CBD agreement released on December 16, 2016 "merely urged caution in field-testing the products of synthetic biology, including gene drives, while supporting better risk-assessment of the products' potential effects" (Callaway 2016).

Although not mentioned by the National Academies of Sciences, Engineering, and Medicine report, three additional international bodies could potentially regulate gene drives: the Convention on the Prohibition of Military or Any Other Hostile Use of Environmental Modification Techniques (ENMOD 1978), the Convention on the Prohibition of the Development, Production and Stockpiling of Bacteriological (Biological) and Toxin Weapons and on their Destruction (BTWC 1975), and the United Nations Committee on World Food Security (see also Thomas 2016; Chapter 5, this volume). This brings me to consider the legalities of CRISPR more broadly: the role of law in, and its relationship with, technology.

CRISPR Legalities

Although still relatively new, CRISPR has already engendered important social questions that are at the heart of our relationship with nature, with the world, with our foods, and of course with ourselves. This relationship is filtered through, and in many cases determined by, legal institutions, legal procedures, and legal norms, which are in turn also impacted by the new technological challenges. In this sense, CRISPR, gene drives, and other gene editing technologies are not only enabled and governed by law, but also constitute regulatory platforms, a mutual process that Sheila Jasanoff refers to as "co-production" (2014).

Consider food as an example. In April 2016, the United States Department of Agriculture (USDA) announced that a CRISPR-edited mushroom, which was modified to not turn brown, did not need to receive the agency's approval before being released to the market (Waltz 2016). This was because genetic material had been deleted, not added, and because the mushroom did not contain foreign DNA from "plant pests" such as viruses or bacteria. These organisms were used for genetically modifying plants in the 1980s and 1990s, when the United States government developed its framework for regulating genetically modified organisms (ibid.). In effect, one might suggest that the mushroom is an outlaw: it stands outside of the law. As such, it illuminates the existing regulatory assumptions that only unnatural additions constitute a "regulatable" change. In this way, not only does the mushroom's classification as a fungus rather than an animal, and its proposed alteration as natural rather than unnatural, impact its regulatory supervision by the USDA rather than by the Food and Drug Administration (FDA), but the law also allows the mushroom population to indeed mushroom as it applies no legal restrictions on its biological life.

Nowhere else does the distinction between natural and artificial perform such a defining role and carry such high stakes in juridical discourse than in the field of gene patent law. After decades of enabling the genetic patenting of any invention that constitutes a novel, useful, and non-obvious patentable subject matter, in 2013 the Supreme Court ruled that "naturally occurring" DNA sequences are not patentable (*Association for Molecular Pathology v. Myriad Genetics, Inc.* 2013). In the language of the *Myriad* court: "We merely hold that genes and the information they encode are not patent eligible under §101 simply because they have been isolated from the surrounding genetic material." Similarly, the Supreme Court ruled in its 2012 *Mayo* case that "laws of nature, natural phenomena, and abstract ideas are basic tools of scientific and technological work that lie beyond the domain of patent protection" (*Mayo Collaborative Services v. Prometheus Labs.* 2012). The Supreme Court's all-or-nothing approach toward nature and technology arguably fails to take into account the complexity, variability, and versatility of these concepts (Sherkow 2014, 1142).

The realm of patent law also sheds light on the problems with the public-private distinction and with the distinction between science and industry. The Broad Institute currently holds dozens of patents on different CRISPR applications. A bitter legal dispute between the Broad Institute, on the one hand, and the University of California, Berkeley and the University of Vienna, on the other hand, over the validity of these patents is awaiting decision by the U.S. Patent and Trademark Office—a decision that will impact the future use of CRISPR by researchers and corporations (Sherkow 2016b; Sherkow and Greely 2015).

In the meantime, on September 22, 2016, the Broad Institute of MIT and Harvard University granted Monsanto a global licensing agreement for the non-exclusive use of CRISPR-Cas9 in agriculture (Monsanto 2016). This agreement, it should be noted, explicitly denies Monsanto the right to use CRISPR for gene drives. Documenting the increasing role of research institutions in "shepherding their researchers' projects through the commercialization process" (Sherkow 2016a, 173), patent law professor Jacob Sherkow cautions against the poisonous effects of commercialization on inter-institutional collaboration, recommending that profitable actions such as granting exclusive licenses or receiving equity ownership in researchers' start-ups should be left to industry.

On the margins of the massive legal disputes over patents, privatization, and the common good, a very different use of patents has also emerged. This time, patents seem to serve as a potent instrument against commercialization and for resisting privatization forces. This was the underlying intention in developmental and evolutionary biologist Stuart Newman's attempt in the late 1990s to patent human-animal chimeras. Rather than privatizing the chimera and reaping profit from it, Newman used the patent application as a performance, devised to explore the social implications of research that scientists prefer to portray as innocuous, sometimes misleadingly so, in his view, and often with the collusion of bioethicists and the media (Chapter 7, this volume).

In his contribution to this volume, entitled *"Sex, Lies, and Genetic Engineering: Why We Must (But Won't) Ban Human Embryo Modification,"* Newman methodically outlines his grave concerns about the current direction of biotechnology. He specifically critiques the commodifying project of making inheritable modifications of human or part-human embryos. Since manipulations of human embryos can have scientific, medical, and commercial benefits, Newman argues, there are powerful incentives to pursue them to increasingly later stages of development. The results will not always be the desired ones, he cautions, concluding that the only way to protect the "humanness" of our species is to ban embryo manipulation in its entirety (Chapter 7, this volume).

Contemplating our ethical and legal stance toward human-animal chimeras and the genetic editing of embryos more broadly is increasingly important as these enterprises emerge from the realm of science-fiction thrillers into real life. In fact, just a couple of days before submitting this text to the press, a long list of scientific journals and media outlets announced a double breakthrough in this field: one team of biologists from the Universities of Tokyo and Stanford has reversed diabetes in mice by inserting pancreas glands composed of mouse cells that were grown in a rat; another group, from the Salk Institute, has shown for the first time that human stem cells can contribute to forming the tissues of a pig (Wade 2017).

Twenty years after Newman's patent application, genetic engineer Kevin M. Esvelt applied for a patent for the CRISPR-gene drive invention. In my extensive interviews with him, Esvelt pointed out that because he was the first scientist to open up the "black box" of gene drives, and since he recognizes the extreme dangers of this technology, he feels responsible to protect the public from its dangers (Chapter 3, this volume). In his contribution to this volume, entitled *"Rules for Sculpting Ecosystems: Gene Drives and Responsive Science"* (Chapter 1), Esvelt outlines a comprehensive program for ensuring that gene drive science is safe and responsive to the public. One of the ways he plans to do so is by holding on to the power to grant licenses for any use of this technology (page 30). Although he critiques the current legal system for its failure to adequately regulate gene editing research, Esvelt feels that he may as well use the "master's tools" to guard the public from inappropriate uses of gene drive technology.

While the circumstances and intentions of their patent applications are admittedly different, both Newman and Esvelt have arguably made subversive use of a highly contentious patent system to protect the public from the perils of what they see as problematic private interests. Stephen Hilgartner considers Esvelt's proposal for the "radical reordering" of science and wonders if it will survive the anachronistic social order within which contemporary scientific and regulatory institutions are embedded. Given the current institutional reality, he stipulates, "efforts to implement Esvelt's vision will likely face an uphill battle" (Afterword, page 193).

In her contribution to this volume, entitled *"Controlling Our 'Nature': Gene Editing in Law and in the Arts"* (Chapter 6), legal scholar Lori Andrews turns her attention from scientists and legal actors to an unconventional group—life science artists—to consider how to further challenge the current regulatory system, "or lack thereof," in the context of gene editing technologies (page 127). According to Andrews, "life science artists predict, reflect, and influence the public's concerns and expectations about genetic technologies, sometimes before the technologies themselves are adopted into society" (page 115). By employing scientific tools for artistic purposes, Andrews argues, life science artists have also raised profound ethical issues about commercialization in the realm of the life sciences and about what can be done so that scientists do not hide their mistakes. She writes:

> When the California Supreme Court in *Moore v. Regents of the University of California* held that John Moore had no property right to his own cell line (which had been secretly patented by his doctor), artist Larry Miller expressed his dismay. Miller, struck by the questions of control and ownership of the body raised by the case, created a Genetic Code Copyright, an elegantly drawn certificate stating: "I . . . born a natural born human being . . . do hereby forever copyright my unique genetic code, however it may be scientifically determined, described or otherwise expressed." He thus challenged the idea that a person can be treated as an object—copyrighted, commodified, and patented.
>
> (Chapter 6, page 122, citations omitted)

Alongside patent laws, the broader legal landscape that governs gene editing technologies is inconsistent and messy, to say the least (Charo and Greely 2015). Although the messiness and inconsistencies seem to be typical of gene editing governance around the globe (Sherkow and Greely 2015, 167–168), this volume will focus mainly on the regulation of gene editing technologies in the United States. Several scholars have already pointed out the incongruities of applying separate legal principles and administration systems to plants and animals (Bergeson 2015; Carroll and Charo 2015; Janis 1991), a typology that has produced awkward consequences in the field. For example, although the United Kingdom-based company Oxitec uses two very similar genomic technologies to reduce population sizes of moths and mosquitoes in the wild, these technologies have been reviewed by two different governmental agencies in the United States: the engineered moth is being reviewed by the USDA, while the engineered mosquito is being reviewed by the FDA and treated as an animal drug (21 U.S.C. Sec. 321(G)(1)(C)) (see also Chapter 5, this volume). As such, the mosquito is also subject to local nuisance and health, safety, and welfare controls.

Acutely aware of the patchy regulatory regime pertaining to gene editing in the United States, the National Academies of Sciences, Engineering, and Medicine cautioned in its June 2016 report about the dangers of having no effective legal mechanism to oversee gene drive research and field trials (2016, 7). The report highlights the unique ability of affected organisms to cross national borders and the inherent property of gene drives to go viral, so to speak. Indeed, mosquitoes and similar insects can fly freely between different jurisdictions, which means that the scientists who are trying to contain these organisms and to protect other organisms from them will need to build the laws of containment into the living organism herself (Chapter 4, page 79).

The idea of containment is the focus of historian of science J. Benjamin Hurlbut's contribution to this volume, entitled "*Laws of Containment: Control without Limits in the New Biology*" (Chapter 4). In his contribution, Hurlbut recounts how anxieties about risk and safety raised by the discovery of recombinant DNA in the 1970s were allayed by the promise of biological containment: the idea that the design of laboratory conditions could ensure that if modified organisms were to escape, they would not survive. Hurlbut explores how biological containment is deployed in the contemporary governance of the biosciences to limit risk and to circumscribe the scope of deliberation, and how it is simultaneously founded upon and underpinned by the notion that biological life is thoroughly controllable, and thus governable. Hurlbut clarifies:

> Approached in this way, biological containment becomes the warrant for setting the New Biology free and for integrating biotechnology into the sociotechnical order: if risks have been excised from the nature of the entity itself, there is no justification for limiting the (presumptively) beneficial development of a technology, or its deployment into the wider world.
>
> (Chapter 4, page 91)

The AquAdvantage salmon offers an illustrative example of this New Biology. In April 2016, the FDA authorized the production of AquAdvantage salmon—the first genetically engineered animal designed for human consumption—and designated it an "animal drug" application (Meghani 2014). Although some have cautioned that engineered salmon could pose a significant environmental risk to their wild relatives, "good governance" has been designed into the salmon in the form of biological containment: all the fish are female, and all are sterile (Chapter 4, page 88). Presented as a revolution in aquaculture, the engineered salmon purportedly offers "abundance without risk" (ibid.).

Here, again, the natural-artificial distinction raises its head (or fin) and is thus policed through scientific means, supposedly establishing and maintaining clear boundaries between wild and human-created life to ensure that the artificial "product" can safely circulate to global markets without posing a risk to natural populations. Then there are also the more traditional physical and environmental forms of containment: the engineered salmon are stocked in artificial ponds in two sites, one in Canada and the other in Panama. They are situated on mountains, hundreds of miles away from freshwater, making the escape from Alcatraz look like getting out of bed in the morning. Finally, the multiple layers of security include "on-facility living quarters for security personnel, security cameras, and 8-foot chain link fencing around each property" (Bodnar 2015). Containment thus turns on a logic of clear demarcation between wild and artificial geographies.

The distinction between natural and human-made has also been important for choosing, and then justifying, certain genetic editing interventions over others. In a lecture to my Law and Genetics class in November 2016, the director of the transgenic company Recombinetics argued that because specific biotechnological applications use natural allele mutations, they are not any different from traditional breeding procedures and should therefore not be subject to any form of regulation, just as human reproduction is largely unregulated. He complained about the popular perception of the DNA as pure and natural, stating that this puritan ideal "is absurd as in fact the genome is a sloppy mess and moves between species" (Fahrenkrug 2016). This argument is quite common among genetic engineers, who stress that breeding crops and animals is an ancient practice that has been managed by Man for centuries. New gene editing technologies allow for the same practices to happen, they argue, only more efficiently ("breeding-on-steroids") and, purportedly, also with less suffering for the individual animals involved (but see Chapter 6, this volume).

In his contribution for this volume, entitled "*Domestic Dogs, Gene Repair, and the 'One Health' Approach*" (Chapter 8), veterinarian Alexander J. Travis offers a similar argument in the context of dog breeding. Specifically, Travis makes a distinction between gene editing practices at large and what he defines as gene repairs (namely, the replacement of a defective gene with a functional copy originating from other individuals within that same species). Gene repair is only different from selective breeding in the precision of the genetic change made, and thus in the greatly reduced number of animals and generations needed to accomplish that change, Travis argues. And because gene repair in dogs could provide significant medical benefits for both dogs and humans, he sees himself as ethically obligated by the Veterinarian's Oath he has taken to use CRISPR-Cas9 for gene repair purposes (Chapter 8, page 158). "This is one of the tenets of the One Health paradigm, which stresses the interconnectedness of all life," Travis concludes (ibid., page 163).

Yet for gene drive scientist Kevin Esvelt, the nature-culture divide does not present an ethical boundary. Quite the contrary, this divide serves as an invitation—a moral obligation even—to engineer. Underlying his pro-engineering approach, I argue in my own contribution to this volume—entitled "*Gene Drives, Nature, Governance: An Ethnographic Perspective*" (Chapter 3)—is Esvelt's more fundamental view of nature at large as "red in tooth and claw." From his standpoint, which is not necessarily shared by the other scientists I interviewed for my chapter, humans are not only allowed, but are even obliged, to, "improve on evolution," and, as long as they do so transparently and responsively, the sky is the limit (pages 28-36).

Because most research is not transparent, however, Esvelt calls for more stringent government regulation of gene drive technology to ensure effective oversight. But until this happens, self-regulation by scientists seems to be the name of the game. If we would like to understand the current governance of novel gene editing technologies, then, we must turn our attention to the scientists who are performing this self-regulation, which is precisely what my chapter offers to do.

I argue, furthermore, that the emotional relationship of these scientists toward nonhumans, nature, and the environment critically informs such self-governance and should thus be explored (ibid.).

Law and Science: An Open-Ended Conclusion

Over the past decade or more, scholars of Science and Technology Studies (STS) have been examining the relationship between law and science, broadly construed. According to Sheila Jasanoff, who participated in the workshop that culminated in this volume: "In all of its guises, actual or aspirational, technology functions as an instrument of governance" (2016, 8). Echoing Jasanoff, several of this volume's contributors explore the ways in which "technology . . . rules us as much as laws do" (ibid., 9; see Chapters 4 and 9 in particular). There is probably no context in which this realization rings truer or more relevant than gene editing, which arguably shapes "not only the physical world but also the ethical, legal and social environments in which we live and act, [enabling] some activities while rendering others difficult or impossible" (ibid., 9). Indeed, biotechnology systems "rival legal constitutions in their power to order and govern society" (ibid.).

Yet despite such realizations about the imbrication of science and law, disagreements about primacy still take center stage in conversations between scientists and social science scholars, as was exemplified in the workshop. For example, Esvelt's not atypical statement that slowly evolving law lags behind the fast pace of science (see also Chapter 1) greatly aggravated the STS scholars in the room. When it comes to science and technology, Jasanoff wrote in this context, the courts are not limited to acting retrospectively, but often lead the way (1995, 11–12). In her words: "the law today not only interprets the social impacts of science and technology but also constructs the very environment in which science and technology come to have meaning, utility, and force" (ibid., 16). Underlying this disagreement about primacy and pace are deep differences in perception about the definition of science and its relationship to governance. Such differences are expressed by this volume's contributors, whose views on the myriad legal, political, ethical, and environmental questions that arise with respect to new gene editing technologies diverge on many fronts—making this collection all the more nuanced and provocative.

Despite the differences and disagreements among this volume's contributors on a variety of issues at the heart of the science-law nexus, all agree on the inadequacy of the current state of gene editing governance. The contributors of this volume also recognize that science today needs to be approached as an instrument of government. And yet, they also realize that despite this recognition, as Jasanoff aptly put it: "there is no systematic body of thought, comparable to centuries of legal and political theory, to articulate the principles by which technologies are empowered to rule us" (ibid., 9–10).

Gene Editing, Law, and the Environment offers one small contribution to the goal of producing such a body of thought. A core premise of this volume is that once we start seeing science as something that governs and is governed by law, then the critical enterprise becomes less about deciding whether or not technology impacts and is impacted by law, and more about realizing the ways in which it is co-produced and co-productive with the law. This collection is thus an effort to bring together a diverse range of scholars and stakeholders to continue the work of examining how technologies are empowered to rule humans and nonhumans in this gene editing era, and by which principles they operate.

Thinking along these lines, anthropologist Gaymon Bennett explores the "social imaginaries" underlying contemporary biotechnology projects. In Chapter 9 of this volume—entitled *"Digital Enchantment: Life and the Future of Gene Editing"*—Bennett contemplates the governing assumptions promoted by scientists and engineers who are bringing gene editing into the contemporary world. Making nuanced observations about the contemporary transition from synthetic biology to digital biology, and situating gene editing within these paradigms, Bennett allows us a glimpse into the hopes and anxieties of the founders of biotechnology and a historical context from which to understand their modes of reasoning. Bennett's insistence that digital biology has recently entered into a zone of "re-enchantment" joins Sandler's "genetic turn," Kuiken's "vigilante environmentalism," and Hurlbut's "New Biology" in an attempt to articulate the novel nature of this emerging era of gene editing life beyond the human.

It is difficult to write words of conclusion that would properly seal this introduction to a volume whose topic is both immediate and constantly in flux. For the contributors of this volume, CRISPR, gene drives, and other leading gene editing technologies provide an entry point to discussing the entangled relationship between law, technology, and our natural and not-so-natural bodies and environments. The importance of this discussion cannot be exaggerated. In the language of the *New Yorker*:

> Pretty soon, we are going to have to make some of the most pressing decisions we have ever made about how, whether, and when to deploy a new technology. The science and some of the researchers may be ready, but society clearly is not, and these decisions are far too consequential to be left to scientists alone.
>
> (Specter 2015)

While some of the contributors would likely question the first part of this statement—i.e., about science always being ready and society always lagging behind—all of us would surely agree on the second part of this statement—namely, that decisions pertaining to the regulation of gene editing technologies are far too consequential to be left to scientists alone.

Acknowledgments

I would like to thank Kevin Esvelt, David Delaney, Guyora Binder, Gaymon Bennett, Ben Hurlbut, Jeantine Lunshof, Jake Sherkow, Stuart Newman, Jack Schlegel, Colin Perrin, Lori Andrews, and Gregor Harvey for their help with this introduction.

References

Association for Molecular Pathology v. Myriad Genetics, Inc. 580 U.S. __ (2013).

Bergeson, Lynn. 2015. *The DNA of the U.S. Regulatory System: Are We Getting It Right for Synthetic Biology?* Washington, DC: Woodrow Wilson Center. October. Available at: www.synbioproject. org/site/assets/files/1388/synbio_reg_report_final.pdf

Bodnar, Anastasia. 2015. "Preventing Escape of GMO Salmon." *Biology Fortified.* November 20. Available at: https://www.biofortified.org/2015/11/gmo-salmon

Brown, Kristen V. 2017. "Why the Patent Battle Over CRISPR Matters (And Why It Doesn't)." *Gizmodo.* January 9. Available at: http://gizmodo.com/will-the-battle-over-the-biotech-discovery-of-the-centu-1790981150

BTWC. 1975. "The Convention on the Prohibition of the Development, Production and Stockpiling of Bacteriological (Biological) and Toxin Weapons and on their Destruction." Biological and Toxin Weapons Convention, London, UK, Moscow, USSR, and Washington D.C., U.S.

Callaway, Ewen. 2016. "'Gene Drive' Moratorium Shot Down at UN Biodiversity Meeting." *Nature News.* December 21. DOI: 10.1038/nature.2016.21216

Carroll, Dana and Alta R. Charo. 2015. "The Societal Opportunities and Challenges of Genome Editing." *Genome Biology* 16: 242–251.

Charo, R. Alta and Henry T. Greely. 2015. "CRISPR Critters and CRISPR Cracks." *The American Journal of Bioethics* 15: 11–17.

Cohen, Jon. 2016. "'Any Idiot Can Do It.' Genome Editor CRISPR Could Put Mutant Mice in Everyone's Reach." *Science Magazine News.* November 3. Available at: www. sciencemag.org/news/2016/11/any-idiot-can-do-it-genome-editor-crispr-could-put-mutant-mice-everyones-reach

DARPA. 2016. "Setting a Safe Course for Gene Editing Research." Defense Advanced Research Projects Agency, September 7. Available at: www.darpa.mil/news-events/2016-09-07

ENMOD. 1978. "Convention on the Prohibition of Military or Any Other Hostile Use of Environmental Modification Techniques." Environmental Modification Convention, Geneva, Switzerland.

Esvelt, Kevin et al. 2014. "Concerning RNA-Guided Gene Drives for the Alteration of Wild Populations." *eLife* 1: 3e03401.

Fahrenkrug, Scott. 2016. Founder and Director, Recombinetics. Skype communication, November 21, 2016.

Friends of the Earth. 2016. "Genetic 'Extinction' Technology Rejected by International Group of Scientists, Conservationists and Environmental Advocates." News Release, September 1. Available at: www.foe.org/news/news-releases/2016-08-genetic-extinction-technology-rejected-by-international-group-of-scientists

Janis, Mark D. 1991. "Sustainable Agriculture, Patent Rights, and Plant Innovation." *Indiana Journal of Global Legal Studies* 9 (1): 91–117.

Jasanoff, Sheila. 1995. *Science at the Bar: Law, Science, and Technology in America.* Cambridge, MA: Harvard University Press.

———. 2003. "In a Constitutional Moment: Science and Social Order at the Millennium." In *Social Studies of Science and Technology: Looking Back, Ahead; Yearbook of the Sociology of the Sciences.* Edited by Bernward Joerges and Helga Nowotny, 155–180. Dordrecht, Netherlands: Kluwer Academic.

———. 2014. *States of Knowledge: The Co-Production of Science and Social Order.* London, New York: Routledge.

———. 2016. *The Ethics of Invention: Technology and the Human Future.* New York: W.W. Norton & Company.

Jasanoff, Sheila, J. Benjamin Hurlbut, and Krishanu Saha. 2015. "CRISPR Democracy: Gene Editing and the Need for Inclusive Deliberation." *Issues in Science and Technology* 32: 25–32.

Kahn, Jennifer. 2015. "The Crispr Quandary." *The New York Times Magazine.* November 9, https://www.nytimes.com/2015/11/15/magazine/the-crispr-quandary.html

Mayo Collaborative Services v. Prometheus Labs. 566 U.S. __ (2012).

Meghani, Zahra. 2014. "Risk Assessment of Genetically Modified Food and Neoliberalism: An Argument for Democratizing the Regulatory Review Protocol of the Food and Drug Administration." *Journal of Agricultural and Environmental Ethics* 27: 967–989.

Monsanto. 2016. "Monsanto Announces Global Licensing Agreement with Broad Institute on Key Genome-Editing Application." September 22. Available at: http://news.monsanto.com/press-release/corporate/monsanto-announces-global-licensing-agreement-broad-institute-key-genome-edi

National Academies of Sciences, Engineering, and Medicine. 2016. *Gene Drives on the Horizon: Advancing Science, Navigating Uncertainty, and Aligning Research with Public Values.* Washington, DC: The National Academies Press.

Reardon, Sara. 2015. "Gene-Editing Record Smashed in Pigs." *Nature News.* October 6. Available at: www.nature.com/news/gene-editing-record-smashed-in-pigs-1.18525

Regalado, Antonio. 2015a. "DuPont Predicts CRISPR Plants on Dinner Plates in Five Years." *MIT Technology Review.* October 8. Available at: https://www.technologyreview.com/s/542311/dupont-predicts-crispr-plants-on-dinner-plates-in-five-years/

———. 2015b. "First Gene-Edited Dogs Reported in China." *MIT Technology Review.* October 19. Available at: https://www.technologyreview.com/s/542616/first-gene-edited-dogs-reported-in-china

Sherkow, Jacob S. 2014. "The Natural Complexity of Patent Eligibility." *Iowa Law Review* 99: 1137–1196.

———. 2016a. "Pursuit of Profit Poisons Collaboration." *Nature* 532: 172–173.

———. 2016b. "Who Owns Gene Editing? Patents in the Time of CRISPR." *Biochemist* 38: 26–29.

Sherkow, Jacob S. and Henry T. Greely. 2015. "The History of Patenting Genetic Material." *Annual Review of Genetics* 49: 161–182.

Specter, Michael. 2015. "The Gene Hackers." *The New Yorker.* November 16. Available at: www.newyorker.com/magazine/2015/11/16/the-gene-hackers

———. 2017. "Rewriting the Code of Life." *The New Yorker.* January 2. Available at: www.newyorker.com/magazine/2017/01/02/rewriting-the-code-of-life

Thomas, Jim. 2016. "The National Academies' Gene Drive Study Has Ignored Important and Obvious Issues." *The Guardian.* June 9. Available at: https://www.theguardian.com/science/political-science/2016/jun/09/the-national-academies-gene-drive-study-has-ignored-important-and-obvious-issues

Travis, John. 2015. "Making the Cut." *Science* 350: 1456–1457.

Université Paris-SUD 11. 2017. "CRISPRs web server." Last updated January 2. Available at: http://crispr.i2bc.paris-saclay.fr

U.S. White House. 2016. PCAST Letter to the President on Action Needed to Protect Against Biological Attack. November 15. Available at: https://www.whitehouse.gov/sites/default/files/microsites/ostp/PCAST/pcast_biodefense_letter_report_final.pdf

Wade, Nicholas. 2017. "New Prospects for Growing Human Replacement Organs in Animals." *New York Times.* January 26. Available at: https://www.nytimes.com/2017/01/26/science/chimera-stemcells-organs.html?hp&action=click&pgtype=Homepage&clickSource=story-heading&module=second-column-region&region=top-news&WT.nav=top-news

Waltz, Emily. 2016. "Gene-Edited CRISPR Mushroom Escapes US Regulation." April 14. Available at: http://www.nature.com/news/gene-edited-crispr-mushroom-escapes-us-regulation-1.19754

Wang, Yanpeng, et al. 2014. "Simultaneous Editing of Three Homoeoalleles in Hexaploid Bread Wheat Confers Heritable Resistance to Powdery Mildew." *Nature Biotechnology* 32: 947–951.

Whitworth, Kristin M. et al. 2016. "Gene-Edited Pigs Are Protected from Porcine Reproductive and Respiratory Syndrome Virus." *Nature Biotechnology* 34: 20–22.

Wu, Jun et al. 2017. "Interspecies Chimerism with Mammalian Pluripotent Stem Cells." *Cell* 168: 473–486.e15.

Part I

Conserving Nature, Driving Evolution

Chapter 1

Rules for Sculpting Ecosystems
Gene Drives and Responsive Science

Kevin M. Esvelt

Introduction: A Relatively Accessible Technology that Allows One to Affect Many

Few dog owners believe that our beloved companions would be able to out-compete the wolves in Yellowstone. No matter how impressed we are by our canine companions, seeing a wolf pack in the wild makes the idea seem absurd. Even so, the dog-to-wolf comparison is representative of a general rule: there is little need to worry about engineered organisms surviving and proliferating in the wild. As Darwin put it, "Man selects for his own good, Nature for that of the being which she tends" (Darwin 1859).

When we tinker with an organism, whether by selective breeding or precision genome editing, we're making changes to a system that evolved to optimize reproduction in its ancestral habitat. Since our changes are for our own benefit, they are highly unlikely to benefit the organism in that original context, and natural selection would consequently eliminate them. There may be exceptions, especially when we've changed the environment, but the default assumption is that we are not nearly as good as evolution in the wild.

Of course, this assumes that the normal rules of inheritance apply, that each gene has an equal chance of being inherited by offspring. In nature, some genes have evolved to break the rules: they have a better-than-even chance of being inherited. These "gene drives" can spread through populations even if they decrease the organism's ability to reproduce: even though carriers produce fewer offspring, more of them will inherit the gene drive (Burt and Trivers 2009). With the advent of CRISPR genome editing, this is a trick we can duplicate (Esvelt, Smidler, Catteruccia, and Church 2014).

Instead of simply replacing an existing DNA sequence with a new version, we can additionally encode instructions for the cell to perform the same replacement on its own (ibid.). Simply insert genes encoding the CRISPR system and guide RNAs directing it to cut the original sequence next to the edited version. When we introduce this DNA into the germ line cells of an organism, those that will go on to make sperm or eggs, CRISPR will precisely cut the target sequence. The cell repairs the damage by copying our sequence in its place. Once one copy is

Altered Gene

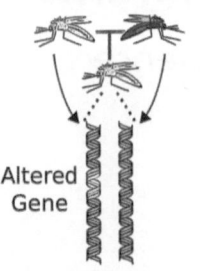

Altered
Gene

1 copy
50% chance of inheritance

CRISPR Gene Drive

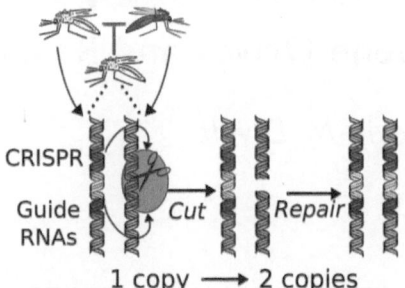

CRISPR

Guide
RNAs

Cut

Repair

1 copy ⟶ 2 copies
100% chance of inheritance

Normal Inheritance

Altered Gene   Wild-Type

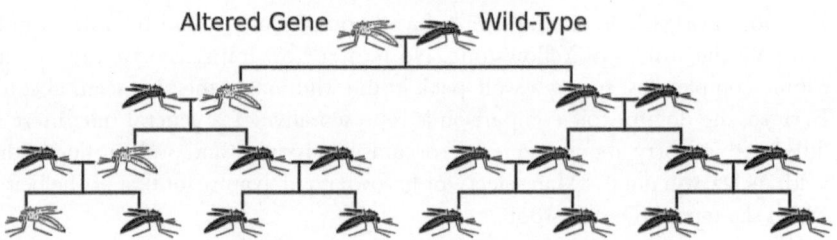

Gene Drive Inheritance

Altered Gene
+ Gene Drive   Wild-Type

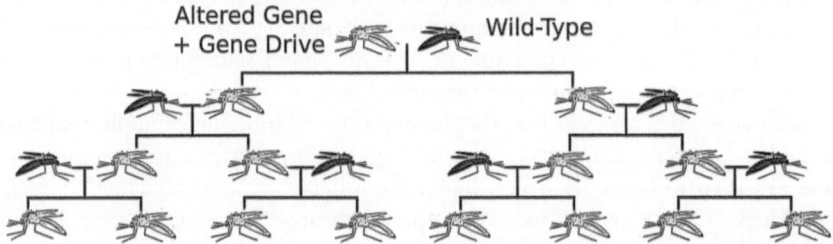

Figure 1 Edited genes in sexually reproducing organisms normally exhibit Mendelian inheritance: when an engineered organism with two copies mates with a wild-type organism, the offspring are guaranteed to inherit one copy. But when these "heterozygotes" mate with a wild-type organism, the offspring have only a 50 percent chance of inheriting the altered gene. If released into the wild, the edited gene will not spread. In contrast, offspring that inherit the edited gene and a CRISPR-based gene drive system exhibit genome editing in their reproductive cells: CRISPR cuts the wild-type sequence, causing the cell to copy the edited gene and the CRISPR components in their place. With two copies, all progeny are guaranteed to inherit the edited gene and the drive system. This process repeats in subsequent generations, causing the edited gene to spread through the population. Such a "global" drive system is self-sustaining: it could potentially spread to every population of the target species in the world.

Courtesy of Kevin M. Esvelt.

inserted, the cell can produce the CRISPR machinery on its own. It cuts and replaces the other copy of the gene, so now there are two. When that organism mates with a wild counterpart, all of the offspring are guaranteed to inherit a copy. And in those offspring, gene editing happens again: one copy becomes two. When those offspring mate with wild equivalents, all of *their* offspring are similarly guaranteed to inherit the edited gene and the gene drive, as will the next generation, and the next, and so on (Burt 2003). Think of it as a find-and-replace for the entire species (Figure 1). What might we accomplish with such a tool?

First, health. The natural world harbors many sexually reproducing species that profoundly harm human health, any of which might be altered or suppressed to block disease transmission. Imagine a world without malaria or schistosomiasis, dengue or yellow fever, and little risk of Lyme disease and other tick-borne infections.

Second, conservation. Invasive species are a major cause of extinctions, particularly on islands. Gene drive systems might be used to suppress or locally eliminate these invasive populations, potentially saving many native species that would otherwise become extinct.

Finally, eco-friendly agriculture. Imagine a world where crops are grown without pesticides, because the pest species have been altered to dislike the crops' taste.

In short, gene drives offer a way to solve ecological problems using biology, not bulldozers and poisons. The technology is far from theoretical: CRISPR-based drive systems have been demonstrated in yeast (DiCarlo et al. 2015), in fruit flies (Gantz and Bier 2015), and in two species of malarial mosquitoes (Gantz et al. 2015; Hammond et al. 2016). However, this capability changes one of the bedrock assumptions undergirding our understanding of the living world. Recall the example of the dogs and wolves: our default expectation for selectively bred or engineered organisms is that natural selection will eliminate the engineered genes. But with a CRISPR-based gene drive system, this changes: we must assume that it will likely spread in the wild. Even if the organism cannot cross oceans on its own, it may do so by hitching a ride. And even if not, the odds are good that someone, somewhere, will move it for their own reasons. A single researcher now has the power to alter ecosystems, making decisions that could—if not countered—affect everyone.

Risks, Human Fallibility, and the Wise Engineering of Complex Systems

What do we do if a gene drive goes wrong? Suppose that someone were to release a drive system accidentally, or without authorization, or even malevolently. Because CRISPR can be programmed to cut nearly any DNA sequence, we can build a second gene drive to override any change spread by an earlier one (Esvelt, Smidler, Catteruccia, and Church 2014). This second gene drive can restore the phenotype, the original traits of the organism, if not yet the exact DNA sequence. But no matter how we alter an ecosystem—whether with physics, chemistry, engineering, or biology—we can't guarantee that it will return to its original state, even if we remove the source of the change. That's true for a forest fire, an oil spill, and a gene drive. Any of them could force it past an ecological tipping point, an S-curve of no return. And because scientists should hold themselves morally responsible for all the consequences of our work, we should carefully consider whether, when, and how to conduct experiments capable of directly impacting the shared environment.

If we decide we do want to engineer very complex systems, how should we go about it?

Rule one: be humble. If we decide to act, we should always aim to make the smallest possible change likely to solve the problem. This is especially true for evolved systems, because natural selection favors evolvability, which is fostered by having many different weak interactions between components. That means changing one component is likely to affect something unintended. Since we can't reliably predict these consequences, it's best to make as small a change as possible.

Rule two: start local. No matter how minimal the change, don't make that change everywhere in the world at once. Try it out in a small and contained area, observe the effects, and then decide whether scaling up is warranted.

The problem with the CRISPR-based gene drive system just described is that it has everything it needs to copy itself. Once released into a susceptible population, it is a self-scaling system that could have global effects. How do you run a field trial of a self-propagating system? Oddly, a National Academies of Sciences, Engineering, and Medicine report on gene drives suggests it is possible, albeit citing guidelines that predated CRISPR-based gene drives (2016). But consider the difficulty. Suppose we were to pick an isolated island for the trial. It must be far enough away that storms cannot carry a fallen tree harboring a pregnant female organism to the mainland; models of much less potent drive systems have suggested that there is a nontrivial chance of spread in the event of a containment breach (Marshall 2009). Any boat or air traffic must be carefully monitored, lest the organism surreptitiously hitch a ride to populations elsewhere. And the island must be placed under military cordon to keep everyone unauthorized away. Why a military cordon? Because people will intervene for fun or profit. Gene drives are of considerable public interest as technologies go, so the field trial is likely to be public knowledge. Genetic engineering is controversial, and it would look very bad if the drive system escaped, meaning some people will have an incentive to release it. Conversely, if the drive system is likely to offer a benefit, there will be a tremendous temptation for people to move it.

An instructive example is rabbit calicivirus, which was deliberately imported and studied under quarantine by Australian scientists as a possible method of controlling the continent's invasive rabbit population. In 1995, the virus escaped quarantine and spread throughout Australia (Schwensow et al. 2014). Remarkably, farmers in New Zealand, which boasts what may be the most strictly enforced biological border-control system in the world, successfully launched an illegal operation to smuggle rabbit calicivirus into their country (O'Hara 2006).

The government was forced to issue a pardon in exchange for information as to the extent of the introduction; many farmers were reported to show no remorse.

The lesson is that people will predictably intervene to foil the best-laid plans—including efforts to prevent organisms from moving past barriers. For a highly newsworthy technology such as gene drives, it is difficult to imagine scenarios in which outsiders do not interfere.

Quarantined field trials will only be safe if we are willing to establish a military cordon around the island and sink any boat that comes near. The quarantine must remain intact until the drive system is no longer capable of spreading—whether because it is a suppression drive that manages to eliminate the local population, or because it's overwritten by a second drive system that cannot spread through unaltered organisms. Such an extreme quarantine, although possible, would be high-risk.

Gene Drive Risks: Primarily Social

Still, what exactly is the nature of the concern? What risks are posed by gene drive systems? Are ecosystems so fragile that an accidental release from a laboratory would cause problems?

The answer is no. Most gene drive systems, especially those that might be accidentally released, are thought to pose very little, if any, ecological risk. It is easy to build a drive system, but it takes careful engineering to build one that can't be blocked by natural DNA sequence variation in the population, or even by the occasional incorrect copying event following CRISPR cutting (Noble et al. 2016a). The first drive system created in fruit flies is costly enough to the organism that mutants of the wild-type sequence—which preserve the original function of the target gene and cannot be cut—will quickly outcompete the drive system. If the drive system were to escape, a wave of yellow fruit flies would likely spread around the globe, quickly followed by a reversion to the usual color, thanks to natural selection favoring the resistant mutants—which are themselves generated by the drive system at a nontrivial rate. Ecological effects are therefore highly unlikely.

What about extinction? There has been a great deal of misleading press about "the extinction invention," written by journalists who focus on the potential to build population suppression drives that spread infertility or bias the population toward one sex (Regalado 2016). But models that take into account drive-resistant mutations clearly show that these will prevent the population from diminishing to the point of no return (Noble et al. 2016a; Marshall et al. 2016). If the goal is the removal of the organism, this will require frequent releases of organisms carrying suppression drives and most likely more than one version. A single unauthorized release is exceedingly unlikely to suffice. Moreover, the process would take dozens of generations, which is more than enough time for anyone so inclined to build organisms that deliberately contain mutations blocking the suppression drive; if desired, these could themselves be spread by their own drive system to immunize the population. Somewhat surprisingly, personal conversations with ecologists have revealed that they are among the most sanguine of scientists when it comes to potential impacts of CRISPR-based gene drive.

That's not to say there is no ecological risk. Rather, the magnitude will differ tremendously by the species and nature of the change. Notably, gene drive systems will directly impact only a single target species, unlike the many other ways we are already impacting ecosystems globally. In the context of the Anthropocene and the sixth great mass extinction (see Chapter 2, this volume), gene drives are small potatoes, especially because phenotypic changes can be overwritten.

The same logic implies few security risks. Again, drive-mediated changes would be slow to spread, requiring many generations of vertical parent-to-offspring transmission. They'd be easily detected by sequencing the genome, because CRISPR gene drives have a uniquely recognizable DNA signature that is probably impossible to hide. And, of course, these changes could be overwritten by a subsequent immunizing reversal gene drive. A weapon that is slow, easily detected, and readily countered does not present a major threat, certainly not relative to many other available technologies.

So what's the problem?

In the modern era, perception is paramount. Suppose one or more organisms with a global drive system escape from a field trial or from a laboratory and are lucky enough to find mates. Because most offspring will inherit the drive system, there's a reasonable chance that there will soon be enough that stochastic chance is unlikely to eliminate the drive. If not overwritten, it will likely spread to every population of that species in the world (Esvelt, Smidler, Catteruccia, and Church 2014).

Imagine the headline: "Scientists accidentally convert an entire wild species to GMOs. Is CRISPR to blame?" The damage to public trust in scientists and governance would be severe and long-lasting. At a minimum, it would be the end of hopes to use gene drives against malaria and schistosomiasis. A mere decade-long delay could keep us from preventing millions of deaths and billions of infections. Judging by past ethical and safety lapses in fields such as gene therapy, a decade would let us off lightly. Add to this situation a dash of widespread public suspicion of genetic engineering, thanks in part to companies such as Monsanto, season it with recurrent media stories every time the drive system spreads to a new city, country, or continent, and bake it with additional incitement from loud anti-technology activists, and you have the recipe for a decades-long delay for gene drives and serious potential damage to more mundane CRISPR-based applications— a very large fraction of all modern biotechnology.

So what can we do? We can raise awareness of the nature of the peril among the scientific community, including the existence of laboratory safeguards that could prevent accidental releases. With this technology, a single lapse could be enough. Since human error is inevitable, it's always best to build the safeguards into the system to account for and mitigate such human error, accidental or deliberate.

Raising awareness is all the more essential because existing biosafety committees and authorities are simply not qualified to evaluate gene drive risks. They are used to looking at antibiotic-resistant bacteria and their possible effects on human health, not at the question of whether an organism will escape and cause ecological risk. Education is vital, but may not be enough.

The policy failure is broader than biosafety committees, however. Existing legal mechanisms cannot adequately regulate new technologies such as gene drives. Since a group of genetic and political scientists (including myself) first called for regulatory reform coincident with our public description of the technology in July 2014, only the Netherlands has crafted new regulation specific to gene drives, which simply amounts a *de facto* ban on research without explicit government permission. No other country has done anything at all. In short, governments appear to be paralyzed in the face of rapidly advancing technologies—not necessarily because their response time is any slower than in the past, but because technologies today advance more quickly. No longer can they take a decade or more to recognize a problem and hash out a response; in many cases, they don't even have a year. Worse, the problem is international. Scientists from dozens of countries have the technical capacity to build CRISPR-based gene drive systems, at least in model organisms such as fruit flies and mice. Once released, these organisms will not recognize borders. The current regulatory brakes are ineffective and inappropriate in the face of this new challenge.

Self-Regulation?

If national laws cannot adapt swiftly enough, might the scientific community regulate itself? The famous 1975 Asilomar meeting was the culmination of this approach, but in practice it suffers from serious drawbacks (see Chapter 4, this volume). The first CRISPR-based gene drive was constructed in yeast after disclosure of the technical possibilities, using two forms of molecular-level confinement to ensure that it could never spread in the wild regardless of human error (DiCarlo et al. 2015). The second was constructed in fruit flies, without using these precautions, by scientists who were trying to do something completely unrelated (Gantz and Bier 2015). Like most others, they hadn't read any of the earlier publications, seen any news coverage, or heard the warnings from other scientists. Indeed, seeing their approach as a laboratory technique rather than a means of harnessing an existing population-level phenomenon, they did not refer to it as a gene drive. In subsequent months, they generously joined my colleagues and me in laying out consensus safeguards for laboratory research (Akbari et al. 2015). But it was a near-miss: publication of their methods without mention of the hazards, which nearly occurred, would quite possibly have led to an accidental gene drive release.

One might argue that the introduction of CRISPR-based gene drive technology could serve as a model for potentially fraught advances: the initial discoverers identified potential problems *before* laboratory demonstrations, consulted with outside experts from many different fields, and ultimately decided to disclose the capabilities and call for regulation before anyone began experiments. For this, I am deeply grateful to my colleagues and mentors, particularly George Church,

Kenneth Oye, Jeantine Lunshof, and James P. Collins. Numerous researchers from diverse fields were aware of the danger of an accidental release, and actively sought to contact other interested laboratories to ensure that others were aware of the existence of necessary safeguards. Were self-policing reliable, all laboratory gene drive experiments to date would have used these strict confinement strategies. Unfortunately, they have not. The culprit is a fundamental problem with the structure of the scientific enterprise: the vast majority of research is conducted behind closed doors.

The Imperative for Open Science

As a rule, researchers in the life sciences emphatically do not share their proposals with competitors. Under the current system, any scientist who discloses a brilliant idea invites competitors to throw more money and hands at it, publish first, and claim all of the credit.

From a societal standpoint, this is senseless. Conducting research blind to the efforts of others is wasteful and inefficient, depriving us of collaborators and forbidding us from making informed decisions on whether to cooperate or compete. No society would rationally design the current closed-door scientific enterprise; both the public and most practicing scientists would almost certainly be better off under an open system. The challenge is making the switch (Esvelt 2016).

In established fields, particularly those where profits from intellectual property are there for the taking, there is little hope of a change. New fields have a better chance, but the logic that everyone would be better off is defeated by the nature of the collective action problem: given current incentives, potential cheaters are better off defecting. We need to start by deliberately working to change the incentives in a field in which other imperatives already favor openness. Starting with one area of research provides a test of the hypothesis that sharing proposals and ongoing research will benefit both scientific progress and safety: if successful, we can scale up by changing the incentives in similar fields, and perhaps eventually to all of science.

Gene drives are an ideal starting point for moral and practical reasons. The moral reasons were perhaps best articulated by the U.S. National Academies of Sciences, Engineering, and Medicine report on gene drive (2016). This is a document that has some flaws: most notably, it uncritically accepts the World Health Organization's advice on the use of field trials to test engineered mosquitoes, even though that document was written before CRISPR-based gene drives came into the picture. Then again, it's unfair and unrealistic to expect any committee to call for radical changes. But even if they were unable to explicitly call for open gene drive research, they included two lines that are pure gold:

> The best course of action is to ensure that the people who could be affected by
> a proposed project or policy have an opportunity to have a voice in decisions

about it. Experts acting alone will not be able to identify or weigh the true costs and benefits of gene drives.

(ibid., 2016)

The first line encapsulates what makes this technology different from others: gene drive can directly impact the shared environment. Indeed, traditional CRISPR-based gene drive systems are global in that they should be assumed to spread to every population in the world; a single accident with the wrong species could spread to ecosystems worldwide. Deciding to perform the experiment is itself a decision that could affect others; doing so in secret explicitly denies these others a voice. Hence, closed-door gene drive research is inherently contrary to societal values.

The second line encapsulates the main flaw of the current scientific enterprise: we are pursuing new technologies in small groups of specialists who cannot reliably anticipate the consequences. Because this work is done behind closed doors, we are actively prohibiting others from helping us predict outcomes. In short, we're searching for and opening technological boxes in the dark. Not because we haven't invented the lightbulb, but because the system effectively punishes anyone who supports flipping the switch. Why should anyone trust us to develop this technology in the dark? The answer is that they probably will not, meaning that openness is likely a prerequisite for deployment.

Thus, the most important application of gene drives may be to engineer the scientific ecosystem (Esvelt 2016). It is the catalyst with which we can demand change from those who control the incentives: scientific journals, funders, policy makers, and holders of intellectual property. Journals determine recognition and status, so any practice required by enough journals becomes a *de facto* requirement for all scientists, particularly if all of the "top" journals sign on.

Recognition also determines funding: without a publication (or possibly a preprint in some fields), it's nearly impossible to acquire enough support to continue research. Hence, if enough funders require a certain practice, it similarly becomes mandatory for all research groups that rely on at least one of those funders. Finally, policy makers can directly determine what groups can or cannot do, but unlike journals and funders, they are limited to governing research within particular nations; few if any international treaties have sufficient enforcement to ensure compliance from researchers. We are actively working to build agreements among journals to require pre-registration of gene drive projects for eventual publication, among funders to require public disclosure of grant proposals involving gene drive, and continuing our efforts to encourage policy changes.

It is important to note that none of these strategies is perfect on its own, but they can be powerful together. For example, at least one research group has informed me that they currently intend to duplicate the earlier fruit fly gene drive experiments without using the recommended safeguards, even though the authors of that original study signed on to a subsequent publication on consensus laboratory safeguards—effectively recommending that no one duplicate their work without taking additional precautions. The justification for such a repetition is simple tradition: duplicate previously reported studies and verify results before making changes to the protocol. Since there are no formal regulations requiring anything else, they will default to this practice. Any kind of formal agreement among journals, funders, or a policy change would dissuade them.

Other situations may be more problematic. Consider a laboratory that has been gifted with secure support from a funder that hasn't agreed to require openness and safeguards. Suppose the financial support is sufficient to meet all their needs for many years, long enough that they are willing to give up their chance at publication in the top journals in order to move their projects forward—for example, they might sincerely care about deployment somewhere in the world that is less likely to care about openness. Finally, suppose that they reside in a nation without specific regulations. Such a research group could freely continue to do their work in secret, including working on the ideas of others, possibly publishing them in lesser journals that did not sign on, or simply releasing their results as preprints. Even one such uncooperative actor could severely undermine the case for continued openness among other researchers in the field.

Intellectual Property as a Regulatory Solution

Unlike government policy and most international treaties, the reach of intellectual property is truly international. Notably, patent law could be mobilized to effectively regulate new technologies more rapidly than government bureaucracies, let alone congressional action. Unlike journals and funders, patent licensing applies even to research groups with secure funding and little need to publish, although the extent of control depends on the nature of the research exemption in the relevant country.

Together with colleagues, I am currently laying the groundwork to use the intellectual property that I filed with Harvard and MIT to regulate CRISPR-based gene drive. Specifically, we aim to create a nonprofit that would take responsibility for issuing research licenses. These would be granted at no cost, but would be conditional upon researchers publicly sharing proposals before experiments begin, providing regular public updates, and abiding by the recommended safeguards. Following the "copyleft" movement, any recipient of a research license must agree to contribute any relevant intellectual property of its own to the nonprofit for research licensing. A possible fifth stipulation would impose a moratorium on developing for-profit applications until 2026, providing enough time to ensure that the first few projects are supported with public or philanthropic funds. Because philanthropic groups will only back efforts to solve the highest priority problems, this would ensure that the first applications focus on obvious societal challenges, hence avoiding popular suspicions of the profit motive.

Crucially, the nonprofit entity could only grant research licenses, permitting the patent holders to profit from their inventions once any ban on for-profit applications expires. This permits the licensing scheme to hold up in court: in the event of any litigation (to be supported by interested philanthropic groups), the patent holders could reasonably argue that ensuring ethical development increases the likelihood of popularly supported deployments, thereby improving the chance of future for-profit applications. Intellectual property therefore represents a possible fourth lever—in addition to journals, funders, and policy—with which we might ensure that gene drive research is conducted in the open with appropriate safeguards.

Daisy Drives: A Safer and More Ethical Technology for Local Control

Even if we do agree to conduct all gene drive research in the open, there is no guarantee that the technology will eventually benefit the world. Suppose that we want to get rid of malaria. In 2015, it caused an estimated 438,000 deaths, 200 million infections, and over 12 billion dollars in economic damage to the poorest regions of the world (WHO 2016).

Now suppose we can engineer a gene drive that causes malarial mosquitoes to be unable to transmit the pathogen (Gantz et al. 2015), or one that suppresses the population to a level where transmission does not occur (Hammond et al. 2016). How are we going to test it? Recall that a release anywhere would likely constitute a release everywhere, if only because people will deliberately spread it. And without testing, how are the countries harboring that mosquito going to decide whether or not they want to use it? It will certainly affect all of them, so presumably they all have to agree. That political agreement will be difficult even among the African nations that are worst off. Diplomatic consensus will be harder to achieve for applications tackling problems that are less severe.

For example, many people have asked whether we should build CRISPR-based drive systems to target the *Aedes* mosquitoes that spread Zika and dengue. The answer is no: those species are present in more than 100 countries, which together house roughly half the world's population. There is no way that a consensus will happen, so it's probably best not to even build them. The existence of such drive systems would only tempt people who see themselves as heroic rebels, but their release would succeed only in damaging popular and political support for other biotechnology applications.

The bottom line is that the larger the numbers of affected people, the harder it will be to win support. Hence, we need a technological way of making locally confined gene drive systems. A drive system capable of spreading alterations through a local population would enable each community to make its own decisions without forcing them onto others—and to make those decisions on the basis of safe field trials, which will be needed because we cannot accurately predict consequences without small-scale experiments in the target environment.

The problem with the standard CRISPR-based gene drives, the reason they are anticipated to spread globally, is that they have everything they need to copy themselves. To solve this, we can split up those elements into a serially dependent "daisy chain" (Noble et al. 2016b). In this daisy chain, element C has the components necessary to cause element B to drive, while B has the machinery needed to drive element A. But nothing allows element C to drive: it is a normal engineered gene, save that it always co-occurs with B and A. Since it will certainly be costly—just about everything is—natural selection will gradually eliminate it. Once C is gone, B no longer drives, and consequently begins to decrease in frequency. Once B is gone, A is a normal engineered gene.

Each link in the daisy chain of a daisy drive is analogous to the booster stage of a rocket: while present, it propels the payload to higher frequency in the local population. Adding more links to the daisy chain increases the extent of the spread, enabling it to be tuned to a particular application in a specific location; other enhancements block gene flow between engineered and wild organisms to make changes correspond to political boundaries.

While still very early in development, daisy drive and other local systems are the future of gene drive technology for technical, social, and ethical reasons. On a technical level, global drive systems cannot be used for conservation: attempt to suppress populations of invasive rats on an island, and they will almost certainly spread the drive system to the mainland. The same is true for agricultural applications to control crop pests. It's better to let farmers make decisions concerning their own land: no single solution will be appropriate for every farm, and the farmer is the local expert. The same logic holds for other problems related to shared ecosystems: local communities know their own environment best.

Socially, agreement is more likely to be reached at a local level. Even for terrible scourges such as malaria, where many people are crying out for a solution, it is far more likely that a single hard-hit village or city will decide to experiment with a local drive system with permission from their national government; should it work well, more communities will follow. Without those stepping-stones, the path to a continent-wide agreement is murky at best.

Perhaps counterintuitively, the larger the number of people who are affected by a technology, the less their voices will be heard in any meaningful way. Scale brings efficiency, but at the cost of precision and democratic legitimacy. By allowing each community to make decisions concerning its own shared environment without forcing those choices on others, daisy drives offer a more ethical approach to altering the natural world.

Where to Start?

The optimal testing grounds for nascent gene drive technologies may be the developed world. This may seem surprising, as there are certainly far more pressing problems in the developing world. But rich-world communities have four main advantages. First, they have strong institutions with long traditions of local decision making. Second, their highly educated populations mean that numerous residents will have relevant expertise and can advise both researchers and their fellow citizens. Third, no one will accuse the project leaders of neocolonialism. And fourth, less severe problems pose fewer ethical trade-offs.

Compare the problem of tick-borne disease to malaria. Over the last few decades, the rapid rise in the incidence of diseases such as Lyme in the northeastern United States (CDC 2015) has nearly destroyed the iconic image of American childhood: kids running freely through the woods. Not unreasonably, many parents in hard-hit communities are reluctant to send their kids outside for fear that they will miss a tick bite, fail to catch the symptoms early, and condemn their child to a life of chronic disease. It is a horrific problem, but it is not killing kids every few minutes. This is in contrast to malaria, where every extra day spent on safety testing and community discussions may effectively condemn 1,000 children to death (WHO 2016). There is no escaping this fundamental ethical trade-off. In the face of a grieving parent demanding immediate action to protect their remaining children, who is prepared to stand and insist on more time for community discussions and democratic legitimacy?

Precisely because it is the most severe problem that could be addressed by gene drive technology, malaria should not be used to develop a general framework for communal decision making about the shared environment.

Preventing Tick-Borne Disease: Open, Community-Responsive Science in New England

In an ideal world, decisions about developing new technologies would be informed by the communities that would be affected. Scientists would publicly lay out the available options, and potential early adopters would make decisions with the help of evaluations of efficacy and side-effects conducted by independent experts. My laboratory and our collaborators are attempting to realize this vision through a project aiming to prevent tick-borne disease in the northeastern United States. To be perfectly clear, the proposal does *not* involve gene drives. Our project builds upon the success of the Eliminate Dengue project (Eliminate Dengue n.d.), which was a pioneer in involving local citizens and running experiments to address their concerns. We have arguably gone even further by inviting potential early-adopter communities to make key technical decisions about the direction of the project at the earliest stages.

Tick-borne disease is an ecological problem. The reforestation of New England without the reintroduction of major predators led to an explosion of the deer population and an equivalent increase in ticks. Because most woods were fragmented by roads and human dwellings, the total amount of woodland-edge habitat is much higher than intact forest, which favors white-footed mice over competing rodents. As it happens, white-footed mice are the best reservoir of Lyme and other tick-borne diseases, much better than other rodents. In short, we have a tick-borne disease problem because we inadvertently engineered the environment to create one (Radolf, Caimano, Stevenson, and Hu 2012).

Together with New England communities, our team is exploring the possibility of heritably immunizing the local white-footed mouse populations responsible for infecting most ticks, and thus for most human infections. If the mice could no longer infect ticks, many fewer ticks would be infected in the next generation. That would equate to fewer infected animals of other species, still fewer infected ticks in the following generation, and so on, effectively disrupting the natural cycle of transmission.

The project began with an initial workshop at MIT in December 2015 that convened molecular biologists, disease ecologists, representatives of environmental NGOs, community leaders, education leaders, physicians, ethicists, and social scientists. The general consensus supported formally contacting potentially interested communities before securing funding or initiating experiments, and ensuring that these communities would be in the driver's seat.

Two potential early-adopter communities stood out. As islands off the coast of Massachusetts, Nantucket and Martha's Vineyard both have long histories of town-hall decision making. Both feature a mix of populations: less affluent year-round residents are joined by large populations of comparatively wealthy and well-educated summer residents. Nantucket is a single polity, while the Vineyard is divided into six separate towns; approaching both of them offers a form of political diversity. Both are in the top few counties for tick-borne disease: the chair of the Nantucket Board of Health estimated that 40 percent of local households had been affected.

We approached the communities to determine whether they were interested enough to begin experiments and to lay out several potential paths forwards (Harmon 2016). There are three technical options for immunization: against Lyme disease only, against tick saliva, or both; each has different risks and benefits. The protective immunity cassette, which would be constructed by isolating the antibody-encoding genes of vaccinated mice, identifying the most protective copies and encoding them in the mouse genome, could be spread by releasing many mice or by using a daisy drive system.

Our laboratory recommends beginning *without* any form of gene drive. Simply put, daisy drive is experimental; we don't yet know if it will work well enough or be stable enough for deployment, and may not learn in time. Either way, initial field trials would be held on an uninhabited island, ideally to be chosen by the communities from among several options. The efficacy and side-effects would be evaluated by an independent group and reported to the communities, which would then make decisions about future directions.

The Inadequacy of the Existing Regulatory Process

Regulations are mandatory, so the tick-borne disease prevention project will certainly follow them. Our team involved regulators at both the federal (FDA) and state (Massachusetts Fish and Wildlife Service) levels from the beginning of the project to ensure we can meet all requirements. However, the current regulatory process is flawed: it is biased towards considering empirical matters, not values, which leaves many citizens feeling ignored. Even more problematic, there is no built-in requirement for community input until the product is already developed and can't easily be changed.

Rather than treating regulatory approval as legitimizing, I would rather treat it as a necessary but insufficient hoop to jump through, and instead work toward and rely on early-stage openness and community guidance to ensure that the technology is developed in an ethical and responsive manner. Ideally, early applications should offer clear benefits to citizens (a greatly reduced risk of tick-borne disease). Discussions should come before experiments, which can ensure that safeguards are agreed upon before laboratory research begins. The project should be developed and run by nonprofit groups, with independent groups monitoring the trials and reporting to the communities (our project is nonprofit, and an independent Data Safety Monitoring Board will perform the analysis of laboratory data and then the field trial). Above all else, all proposals and results should be open, and the project responsive to concerns expressed at the earliest stages (everything has been open from the outset).

While the tick-borne disease prevention project meets what I see as the primary ethical concerns, I harbor no pretensions that it is even close to the best approach to community guidance of research. Indeed, part of the goal is to work with communities to discover how technologies that affect the shared environment ought to be developed. The effort is merely one in a series of steps toward an ever-improving model of how best to solve shared problems, but one that may help encourage gene drive scientists to invite open community involvement at the earliest stages of research.

Conclusion

Gene drive technology highlights several inadequacies of the current legal system and scientific enterprise. Conducting research behind closed doors poses severe moral and practical difficulties when a single experiment can affect many people outside the laboratory, including across international borders, without anyone else taking action. Self-policing by the scientific community has proven unreliable, if only because scientists keep their plans and ongoing work secret from one another, while rapid progress has rendered tentative government attempts at regulation painfully outdated. Even though no one would rationally design the scientific enterprise to favor blind competition as it does, the system evolved before communication technologies provided an alternative and has not yet been able to transition to a new model.

The field of gene drives may offer a controlled means of testing the feasibility and wisdom of open science. Because a single accidental release could prove devastating for public confidence in scientists and governance, ensuring that no such events occur until after the deployment of popularly supported interventions will arguably be necessary for the technology to benefit the world. Using appropriate safeguards and conducting research in the open to help win support will be vital. Toward this end, funders, journals, policy makers, and holders of relevant intellectual property should work to change scientific incentives in favor of open gene drive research. Success will enable technology developers to actively invite concerns and criticism from other experts and especially from early-adopter communities, potentially redesigning the technology at an early stage. Project governance could be informed by ongoing community-guided ecological engineering projects that do not involve gene drives.

By working to change scientific incentives in favor of open gene drive research, pursuing advances that would confer local control over environmental changes, and promoting active involvement by potential early-adopter communities, gene drive researchers can pioneer a faster, safer, more transparent, and community-supported discovery process.

Acknowledgments

I thank Irus Braverman for her helpful comments and suggestions.

References

Akbari, Omar S. et al. 2015. "Safeguarding Gene Drive Experiments in the Laboratory." *Science* 349: 927–929.

Burt, Austin. 2003. "Site-Specific Selfish Genes as Tools for the Control and Genetic Engineering of Natural Populations." *Proceedings of the Royal Society B: Biological Sciences* 270: 921–928.

Burt, Austin and Robert Trivers. 2009. *Genes in Conflict: The Biology of Selfish Genetic Elements*. Cambridge, MA: Harvard University Press.

CDC. 2015. "How Many People Get Lyme Disease?" Centers for Disease Control and Prevention. Available at: https://www.cdc.gov/lyme/stats/humancases.html

Darwin, Charles. 1859. *On the Origin of Species by Means of Natural Selection, Or, The Preservation of Favoured Races in the Struggle for Life*. London: John Murray.

DiCarlo, James E. et al. 2015. "Safeguarding CRISPR-Cas9 Gene Drives in Yeast." *Nature Biotechnology* 33: 1250–1255.

Eliminate Dengue. n.d. "Eliminate Dengue Program." Available at: www.eliminatedengue.com

Esvelt, Kevin M. 2016. "Gene Editing Can Drive Science to Openness." *Nature* 534: 153.

Esvelt, Kevin M., Andrea L. Smidler, Flaminia Catteruccia and George M. Church. 2014. "Concerning RNA-Guided Gene Drives for the Alteration of Wild Populations." *eLife* 3: e03401.

Gantz, Valentino M. and Ethan Bier. 2015. "The Mutagenic Chain Reaction: A Method for Converting Heterozygous to Homozygous Mutations." *Science* 348: 442–444.

Gantz, Valentino M. et al. 2015. "Highly Efficient Cas9-Mediated Gene Drive for Population Modification of the Malaria Vector Mosquito *Anopheles stephensi*." *Proceedings of the National Academy of Sciences* 112: E6736–E6743.

Hammond, Andrew et al. 2016. "A CRISPR-Cas9 Gene Drive System Targeting Female Reproduction in the Malaria Mosquito Vector *Anopheles gambiae*." *Nature Biotechnology* 34: 78–83.

Harmon, Amy. 2016. "Fighting Lyme Disease in the Genes of Nantucket's Mice." *The New York Times*. June 7. Available at: https://www.nytimes.com/2016/06/08/science/ticks-lyme-disease-mice-nantucket.html

Marshall, John M. 2009. "The Effect of Gene Drive on Containment of Transgenic Mosquitoes." *Journal of Theoretical Biology* 258: 250–265.

Marshall, John M. et al. 2016. "Overcoming Evolved Resistance to Population-Suppressing Homing-Based Gene Drives." *bioRxiv*. Preprint, submitted November 17. Available at: http://biorxiv.org/content/early/2016/11/17/088427

National Academies of Sciences, Engineering, and Medicine. 2016. *Gene Drives on the Horizon: Advancing Science, Navigating Uncertainty, and Aligning Research with Public Values*. Washington, D.C.: National Academies Press.

Noble, Charleston et al. 2016a. "Daisy-Chain Gene Drives for the Alteration of Local Populations." *bioRxiv*. Preprint, submitted June 7. Available at: http://biorxiv.org/content/early/2016/06/06/057307

Noble, Charleston et al. 2016b. "Evolutionary Dynamics of CRISPR Gene Drives." *bioRxiv*. Preprint. Available at: http://biorxiv.org/content/early/2016/06/06/057281

O'Hara, Peter. 2006. "The Illegal Introduction of Rabbit Haemorrhagic Disease Virus in New Zealand." *Revue Scientifique et Technique (International Office of Epizootics)* 25: 119–123.

Radolf, Justin D., Melissa J. Caimano, Brian Stevenson and Linden T. Hu. 2012. "Of Ticks, Mice and Men: Understanding the Dual-Host Lifestyle of Lyme Disease Spirochaetes." *Nature Reviews: Microbiology* 10: 87–99.

Regalado, Antonio. 2016. "The Extinction Invention." *MIT Technology Review*, April 13. Available at: https://www.technologyreview.com/s/601213/the-extinction-invention

Schwensow, Nina I. et al. 2014. "Rabbit Haemorrhagic Disease: Virus Persistence and Adaptation in Australia." *Evolutionary Applications* 7: 1056–1067.

WHO. 2016. "World Malaria Report 2015." *World Health Organization*. Available at: www.who.int/malaria/publications/world-malaria-report-2015/en

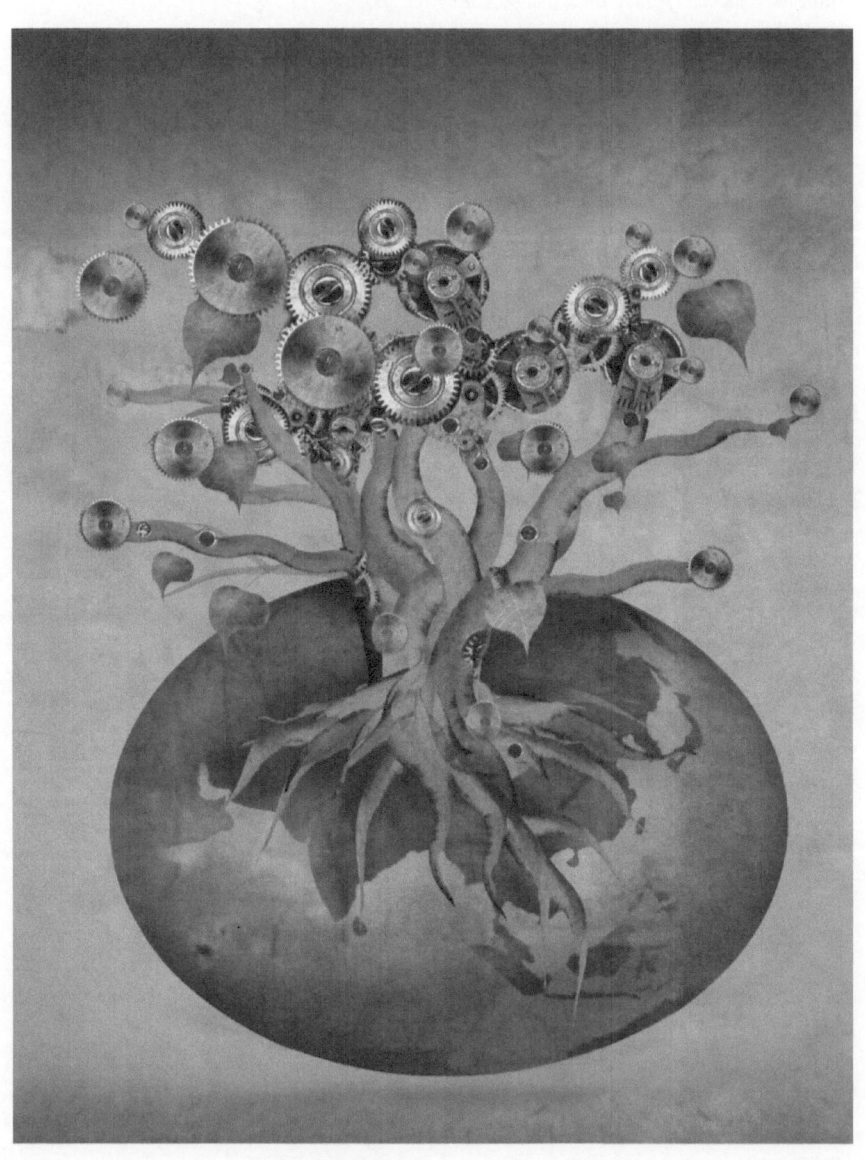

Chapter 2

Gene Drives and Species Conservation

An Ethical Analysis

Ronald Sandler

The ability to drive genetic change through wild populations constitutes a significant increase in the power of humanity (or, more specifically, those with access to the technology) to intentionally engineer ecological systems and communities. It enables new ways of addressing conservation challenges and accomplishing ecological goals. Many members of the conservation and environmental communities believe that we are in desperate need for powerful new tools such as this. As a result, a diverse array of possible environment applications for gene drives has been proposed, including curing diseases, eliminating invasive species, increasing genetic diversity, improving agricultural productivity, and facilitating adaptation to climate change. In this chapter, I conduct an ethical analysis of the use of gene drives in the field for species conservation purposes. The aim is to assess the case for employing gene drives in particular conservation situations, as well as the ways in which this technology may restructure the practice of conservation. Several ethical considerations regarding gene drives intersect with broader discussions about the appropriate roles of human design and engineering in ecological systems. Therefore, this chapter also situates gene drives within the ongoing discourse regarding how we ought to respond to pervasive, large-scale anthropogenic change.

The Conservation Dilemma[1]

Eagerness (or, at least, willingness) on the part of many conservationists to consider novel, more interventionist conservation approaches such as gene drives, assisted colonization, and de-extinction arises from what I have elsewhere called the *conservation dilemma* (Sandler 2012a, 2013a, 2013b). Global climate change, habitat loss and other rapid, and macro-scale ecological changes are dramatically increasing the number of at-risk species. The background or historical rate of extinction is thought to be less than one species per million per year, or 0.000001 percent annually (Baillie, Hilton-Taylor, and Stuart 2004; De Vos et al. 2015; Ceballos et al. 2015). On most estimates, there are 10 to 20 million eukaryotic (plant and animal) species. Thus, a "normal" number of extinctions would be fewer than 20 extinctions per year. However, the current extinction rate is already hundreds, or perhaps even thousands, of times higher than the background rate, due to anthropogenic activities and impacts (De Vos et al. 2015; Ceballos et al. 2015); and global climate change in particular is expected to dramatically increase extinction rates still further. For example, one study found that 24 to 50 percent of bird species, 22 to 44 percent of amphibian species, and 15 to 32 percent of coral species have traits that make them "highly vulnerable" to climate change (Foden et al. 2013). Thus, there is a greater need than ever before for effective conservation tools and strategies for at risk species.

However, global climate change (and large-scale ecological change generally) also undermines the traditionally predominant approaches to species conservation—place-based or *in situ* preservation and restoration. In the increasingly frequent cases where species are threatened by changes in background climatic and ecological conditions, or by deep and irreversible (or difficult to reverse) ecosystem transitions, trying to locally manage spaces to maintain ecological continuity or historical ecological relationships is not feasible (Higgs et al. 2014; Sandler 2012a, 2012b; Minteer and Collins 2010; Camacho et al. 2010; Hoegh-Guldberg et al. 2008). It often will not be possible to conserve species populations in their existing location, or where they have been in the past, when changing ecological and climate conditions are what threaten them, rather than locally manageable stressors, such as poaching or pollution. For example, it is not possible to preserve coral reefs and the species that depend upon them by designating and protecting their locations as marine sanctuaries, when the causes of coral decline are mainly attributed to climate change and ocean acidification due to elevated atmospheric levels of carbon dioxide. It is not possible to preserve American pika populations in the western United States or cloud forest orchid populations in Costa Rica by protecting the mountain tops where they live, when climatically altered temperature and precipitation patterns present the central threats to them. In other words, place-based conservation strategies depend upon the relative stability of background climatic and ecological conditions—and these are in decline.

Thus, the conservation dilemma is that for climate-risk species, either we must move away from species conservation as an ecosystem management goal, or else we must adopt novel conservation strategies that very often involve more intensive human intervention into and manipulation of species populations and ecological systems, thereby putting them in tension with the commitment to preserving historical continuity and human-independent ecological processes. Conservation has traditionally prioritized removing human impacts and refraining from human interference, design, and control. However, species conservation appears to increasingly require greater human intervention. Here is a statement from United States Climate Change Science Program regarding this general tension:

> Over time, some ecosystems may undergo state changes such that managing for resilience will no longer be feasible. In these cases, adapting to climate change would require more than simply changing management practices—it

could require changing management goals. In other words, when climate change has such strong impacts that original management goals are untenable, the prudent course may be to alter the goals. At such a point, it will be necessary to manage for and embrace change.

(United States Climate Change Science Program 2008, Ch. 9, 3)

Here is a similar statement from environmental ethicist Ben Minteer and conservation biologist James Collins regarding species conservation in particular:

The upshot is that we simply have no choice but to think beyond the traditional pars-and-preservation model if we wish to save species in an era of rapid climate change. This will require coming to grips with a significantly more activist and hands-on approach to species conservation than we have taken in the past. It will also mean redeploying our funds and research efforts as we shift them from traditional preservation agendas toward more pragmatic and interventionist programs for conservation science and action on a rapidly changing planet.

(Minteer and Collins 2010, 1802)

When it is no longer possible to accomplish the traditional goals of conservation by the traditional methods, one can either modify the goals (as suggested by the passage above from the Climate Change Science Program) or modify the methods (as suggested in the passage above from Minteer and Collins). Many conservation biologists prioritize species conservation over maintaining independence from humans. They therefore advocate for adopting a more hands-on approach and a more forward-looking perspective with respect to climate–risk species (Camacho et al. 2010; Minteer and Collins 2010; Hoegh-Guldberg et al. 2008; Donlan et al. 2006). According to this view, if conservation managers need to move species to new locations beyond their historical ranges in order to conserve them (i.e., assisted colonization), then they ought to do so, even though it involves intentionally creating vibrant nonnative species populations. If they need to employ more powerful tools to modify ecological system and species populations (i.e., gene drives), then they ought to do so, even though it involves engineering species and systems according to how we believe they ought to be, rather than deferring to human-independent ecological and evolutionary processes.

The Fallacy of the Appeal to the Anthropocene[2]

As discussed above, many proponents of adopting novel species conservation strategies, such as assisted colonization and gene drives, argue for doing so by appealing to the fact that humans have already so modified ecological systems that we now have to take responsibility for shaping and managing them. We must "embrace the Anthropocene," they argue, and doing so means taking a more robust role in designing, engineering, and managing ecological systems, processes, and species.

It is uncontested that human impacts on Earth are already immense. For example, humans appropriate approximately 25 percent of the Earth's primary plant production (Krausmann et al. 2013), over one-third of the terrestrial surface of the Earth is used for agriculture (FAOSTAT 2015), and over 90 percent of global fish stocks are fully or overexploited (FAO 2014). Vertebrate populations are estimated to have been reduced by half on average in the past 40 years due to human activities (WWF 2014) and human activities generate more reactive nitrogen than do all other planetary processes (Galloway et al. 2004). Additionally, the atmospheric concentration of carbon dioxide is higher than it has been in millions of years, primarily due to fossil fuel use, and this is causing the oceans to acidify (NOAA 2015; Tripati, Roberts, and Eagle 2009). Damming, irrigation, channeling, pumping, and floodplain engineering now control or influence the movement of most freshwater and sediment (Syvitski and Kettner 2011). Human activities, particularly agriculture, mining, and building, move more earth than do all other planetary processes (Wilkinson and McElroy 2007). Synthetic chemicals and waste from human industrial activities permeate terrestrial and aquatic systems and organisms. When all of these (and other) impacts are considered together, it is clear that the human influence on the environment is pervasive, transformative, and enduring. The term "Anthropocene" has been widely adopted to refer to this ongoing period of high magnitude anthropogenic planetary change. The fact that there already is such widespread and high magnitude change is frequently taken to countenance interventionist ecosystem management. Given that we are impacting ecological systems so much already, it seems to many that it is better to do so rationally and intentionally, rather than thoughtlessly and recklessly (Marris 2013; Ackerman 2014; Ellis 2009).

However, the inference from the fact that humans have enormous impacts on the planet to the conclusion that we ought to take greater control of ecological systems and processes does not necessarily follow. While many ecological spaces are now novel, no-analog, or hybrid systems, others retain relatively high levels of historical continuity and human independence. Fifteen percent of the terrestrial surface of the Earth is protected and a much larger percentage is ecologically intact and lightly touched by people (UNEP-WCMC and IUCN 2016). This suggests that a unitary response to "the Anthropocene" is not justified. Or, to put this differently, the challenge is not determining how ecosystem managers and conservation biologists ought to respond to "the Anthropocene." The challenge, instead, is in determining how to respond to a wide variety of issues and cases involving anthropogenic change. Appealing to the fact (if it is one) that we are in the Anthropocene does not provide guidance on assisted colonization, ecological restoration, gene drives, or scarce resource allocations, either in general or in particular cases.

Here is a more formal way to put this point (Vucetich, Nelson, and Batavia 2015). It is problematic to jump from this: (1) We are in the Anthropocene; to this: (2) Therefore, we ought to X (where X is any general behavior type). This inference is problematic because either "the Anthropocene" is a strictly descriptive concept, in which case it is invalid to infer from the fact that we are in the Anthropocene (if it is a fact) to a prescriptive conclusion without any additional normative or value premise; or "the Anthropocene" is a normatively loaded concept that smuggles in normativity without sufficient argument for it, in which case the inference commits the fallacy of asserting the conclusion. In either case, the inference is fallacious and the conclusion is not warranted.

Here is yet one more way to think about this. The Anthropocene is consistent with either greater restraint or greater intervention within ecological systems. Merely being in the Anthropocene does not itself tell us which response is warranted. Other considerations need to be provided to settle the issue. Moreover, as just discussed, a blanket position is not likely to be justified. There is too much divergence in types of systems, social contexts, and operative values. In some cases, highly interventionist species conservation strategies might be justified, in other cases more restrained approaches may be. In some cases, highly controlled resource management might be justified, in other cases it might be better to defer to human-independent ecological processes. Therefore, "the Anthropocene" concept may be a distraction when it comes to thinking about novel species conservation and ecosystem management strategies.

Gene Drives as a Conservation Tool

If appeal to the Anthropocene is not an appropriate way to think about whether we ought to adopt more interventionist species conservation strategies such as gene drives, then what criteria ought to be used? As discussed above, large amounts of case specific data will of course be crucial. But there are also several plausible criteria that are commonly appealed to, both explicitly and implicitly, in the environmental ethics and species conservation discourse. All other things being equal, a conservation approach or strategy is generally considered preferable to the extent that it:

- *Sustainably conserves species*: Efforts or strategies that are likely to only forestall extinction are less preferable to ones that are more likely to maintain species in the long run.
- *Addresses the causes of population declines:* Efforts or strategies that address the underlying causes of species population declines are preferable to those that do not, in large part because they are more likely to be sustainable and they remediate anthropogenic ecological harms.
- *Conserves species in their historical ranges*: Efforts or strategies that are able to conserve species *in situ* or in their historical ranges are preferable to conservation *ex situ* or outside of their historical range.
- *Conserves the value of species:* The value of a species population is very often tied to its context. Ecological value, cultural value and natural value are all based

on relational properties between the population and its ecosystem or social context. This is a large part of why *in situ* conservation is preferable.

- *Is scalable and coarse-filter:* The magnitude of the conservation challenge is such that approaches to species conservation that capture large numbers of species or that can be scaled up are preferable to fine-filter approaches that focus on one or a few species at a time.

- *Has public support:* Conservation projects and practices that have strong public support are preferable to those that lack support or face opposition.

- *Has ancillary benefits:* Conservation projects and strategies that have secondary benefits, such as maintaining ecosystem services, promoting ecotourism, or protecting areas that are ecologically important or crucial for other species, are preferable to those that do not.

- *Is feasible and cost-effective:* Conservation funding and other resources—e.g., time and expertise—are finite and scarce. Therefore, approaches that are cost-effective and have a higher probability of success are preferable to those that do not.

- *Has acceptable risks:* Conservation project and strategies that have acceptable and manageable risk—e.g., to ecosystem services, the organisms themselves, or other species—are preferable to those that do not.

These considerations are commonly appealed to in justifying some conservation programs or efforts and in criticizing others. They are all *"ceteris paribus"* (or "all other things being equal") and admit the possibility of exceptions. They are also largely comparative. That is, they evaluate possible conservation efforts against alternative conservation approaches as well as against refraining from conservation activities.

A relevant feature of the criteria for present purposes is that they do not evaluate types of interventions, techniques, or strategies as such. Rather, they help to identify when the use of a particular intervention is well justified. Some instances of assisted colonization will have lower risk, greater public support, and will conserve the value of the target species (e.g., because its value is economic or symbolic) better than will other instances. Moving economically significant conifer species north in Canada has a different conservation profile than does moving polar bears to Antarctica, for example. They are both cases of assisted colonization, but the latter is less likely to succeed, is ecologically reckless, will cause translocated animals to suffer, and would not reestablish the relationships that make the species valuable (see Sandler 2010, for a more comprehensive ethical assessment of assisted colonization).

When evaluating gene drives in this way, they are likely to be assessed favorably in comparison to other possible interventions (including no intervention at all), at least in certain cases. Consider, for example, the proposal to use a gene drive system to skew sex ratios of nonnative rodent—mice and rat—populations on Pacific islands. Islands have extremely high rates of extinctions and endangered species. According to certain estimates, up to 80 percent of recorded extinctions have been on islands, and as much as 40 percent of threatened species are island dwellers (Ricketts et al. 2005). Invasive mice and rats are among the primary drivers of native species declines on islands. They have been widely introduced, leading to dozens of extinctions and hundreds of at-risk species (Atkinson 1985; Doherty et al. 2016) and significantly disrupting ecosystem integrity and functioning. As a result, there have been hundreds of attempts to eradicate invasive rodents from islands, most of them by using a general anticoagulant toxicant (Russell and Holmes 2015). The success rates for such eradication tactics are much higher in temperate than in tropical regions, where it is more difficult to reach all the target animals. Moreover, the widespread application of a general (i.e., not rattus-specific) toxicant can have detrimental impacts on other native species, livestock, or people (Campbell et al. 2015). Therefore, conservationists have a strong interest in developing strategies that are more effective in reaching the entirety of the invasive rodent population and that are target-specific.

Gene drive mechanisms appear to have the potential to address these difficulties and with lower risks than alternative innovations, such as species-specific toxicants or virus-vectored immunocontraception. Thus, some conservationists are interested in developing and deploying them in this conservation capacity by driving sex skewing for maleness through the population. If all offspring are male, then the population will crash in just a few generations. Such an approach would be species-specific and could reach the entire target population. Moreover, deploying gene drives in this way would address the cause of at-risk species declines—i.e., the invasive rodents—and would do so by undoing or removing a human-introduced threat, thereby enabling at-risk species to persist in their native habitats and reestablish historical relationships. Thus, it is an intervention that, if successful, would protect and promote diverse environmental values—namely, ecological integrity, biodiversity, and naturalness.

The foregoing is not intended to be a defense of the use of gene drives for rodent eradications in general, let alone in any particular instances. Rather, it is meant to demonstrate that there are likely to be some cases where gene drive scores favorably on commonly recognized criteria in comparison to the alternatives. Eliminating widespread and difficult-to-target invasive predators or nonnative disease vectors (such as the nonnative mosquito that is spreading avian malaria to endangered Hawaiian birds) appears to be one such type of application. Thus, when thoughtful and responsible researchers begin to develop decision trees, matrixes, and maps for the use of novel genetic conservation interventions there will be branches and pathways that lead to gene drives (National Academies of Sciences, Engineering, and Medicine 2016).

This is not to suggest that the use of gene drives in cases where they are well justified on conservation criteria would not be controversial; they would be and, in some cases, already are. The concerns are likely to include the notion that all forms of intentional gene level engineering are "unnatural" or consist of "playing god" in ways that are intrinsically problematic and therefore wrong (for a discussion of these concerns, and responses to them, see Sandler 2012a). There will also likely be procedural concerns about community consultation and consent, particularly in places with problematic histories of exploitation, colonization, and marginalization of local communities. There will be policy concerns regarding adequate regulation and oversight capacity, as well as risk assessment and management concerns regarding implementation, monitoring, and follow-up. These types of concerns, and many others, are commonly raised regarding interventionist conservation strategies, as well as with regard to the creation and use of transgenic organisms. Gene drives are at the "sweet spot," where these concerns come together. Thus, it is not surprising that many environmental organizations, such as Friends of the Earth, oppose the development of the technology, that many prominent conservationists are critical of it, or that a recent vote related to it at the IUCN's World Conservation Congress resulted in a call for a moratorium on its deployment (and synthetic biology generally) until the technology is further studied (Friends of the Earth 2016). A proposed global moratorium on gene drives was rejected recently at a meeting of the United Nations Convention on Biodiversity (see Chapter 5, this volume).

It is important to note that the most promising and least controversial conservation applications of gene drives are those that would achieve an already widely accepted conservation goal, such as eliminating invasive predators that threaten native biodiversity. This is a common form of justification with respect to conservation genomics—i.e., that adopting them is justified because they would enable conservation biologists to do what they already do, only more effectively. Take, for example, captive breeding programs for endangered species. Analyzing the genomes of the individuals involved in order to increase genetic diversity and avoid inbreeding—which is already done with regard to many endangered species, such as the island fox in California and the owl parrot in New Zealand, for example—simply increases the likelihood of positive outcomes. This, precisely, is the rationale at work in the pursuit of gene drives for eliminating invasive rodents on tropic islands. Conservationists are already attempting to do this; gene drives will simply enable them to do it with greater efficiency and effectiveness.

The most promising cases for gene drives, such as eliminating invasive island predators, are not the hard types of Anthropocene conservation cases, those where shifting ecological background conditions are the major threats. The most promising cases are those in which the source of the threat is local and it is in the power of local managers and actors to address it. Thus, evaluating gene drives as a conservation tool—namely, as merely more effective means to solving a familiar problem—only gets on so far with respect to their ethical evaluation. They must also be evaluated as a novel form of "conservation." This is another reason why appeal to the Anthropocene is not an appropriate way to justify gene drives in general. As the discussion above illustrates, conservation applications of gene drives are often not even a response to the distinctive conservation problems associated with macro-scale anthropogenic change.

Gene Drives and "Conservation"

I have already considered gene drives as a conservation tool that could be employed within a standardly accepted conservation model as a more efficient way to accomplish familiar conservation goals. As such, this technology raises familiar types of concerns associated with risk, control, oversight, process, and consent—and thereby arrogance and hubris—as well as animal welfare and ecological integrity. Most environmental organizations and conservationists concerned about gene drives are critical of them on these grounds. However, in some cases, these concerns could be adequately addressed, so that gene drives can be offered as an attractive conservation option. Therefore, assuming that there is not anything intrinsically problematic, from an ethical standpoint, with intentional genetic modification, there may well be some cases in which it is well justified in comparison with other available options (Sandler 2012a).

At the same time, gene drives also enable an alternative form of "conservation": one that involves the genetic alteration of individuals of a population that is targeted for conservation. Specifically, gene drives could be used to engineer wild population so that they are better fitted or adapted to obtaining ecological, social, and economic conditions. Perhaps panthers can be engineered so that their sex ratios skew slightly female in order to help build back populations (see Chapter 8, this volume); perhaps rhinos can be engineered so that they have stunted horns that would not be valuable to poachers; perhaps frogs and bats can be engineered to be resistant to the fungal diseases that are decimating their populations (chrytridiomycosis and white-nose syndrome, respectively). When gene drives are considered in this way, they can potentially offer solutions to what seem to be intractable conservation problems, including those associated with macro-scale ecological change. Perhaps it is possible to engineer traits into climate-threatened species that will make them more resilient or resistant to the ecological changes that are putting them at risk—e.g., corals that would be able to thrive in more acidic oceans and pikas that are less sensitive to heat.

If deployed in this way—to creatively engineer at-risk populations to be better adapted to obtaining conditions—gene drives would fail to satisfy many of the conservation criteria that make their other applications attractive. For example, the interventions would not address the underlying causes of species declines or reestablish the ecological relationships that are the basis for species value. However, this use of gene drives would enable a new model for conservation. Rather than focusing on the conditions around the target species—e.g., air and water quality, human recreation, invasive species, and habitat quality—they would make it possible to ask: How can we adapt the species so that it is better suited to these conditions? The fix of the ecological problem will be at the level of the species.

This is a new question for conservation biologists (though not for animal and plant breeding). It involves a novel type of conservation perspective and an expanded conception of what counts as conservation. It also connects gene drives back to the issue raised at the beginning of this chapter—namely, whether we ought to take a more engineering-oriented approach toward the natural world—not merely in terms of how conservation is accomplished—the tools and means—but with respect to the ends. Should we try to make species how we believe they ought or need to be, rather than try to conserve them how they are, even if not where they are?

Similar questions have arisen regarding the recovery of ecologically degraded sites. Proponents of ecological restoration have begun to grapple with the reality that historical reference conditions for assisted recoveries are less likely to be reliable guides for future ecological integrity under conditions of high rate and high magnitude anthropogenic change (Hobbes, Higgs, and Harris 2009; Higgs et al. 2014; Harris, Hobbs, Higgs, and Aronson 2006). So assisted ecological recoveries that aim to be very high on historicity—i.e., to return ecological spaces to what they were or would have been absent anthropogenic degradation—can sometimes be a form of ecological insensitivity. However, the fact that it is often not feasible or advisable to try to recreate past ecological trajectories does not necessarily license engineering ecological systems as we believe they ought to be. In addition to trying to accomplish high levels of historical fidelity and intensively engineering systems according to our designs, there is also a third possibility: to largely defer to ecological and evolutionary processes by providing space and time for ecological systems to reconfigure and transition. The results may not be what people would prefer on the time scales that they would prefer them. But, in some cases, this strategy may be more ecologically appropriate, and it might reinstantiate some of the natural and ecological values lost through human degradation that additional human design and control might not (Sandler 2012b). If the goals are ecological and biological, then the time scale and perspective from which management is evaluated may need to be so as well.

The point here is not to oppose creative gene drive applications or highly managed ecological restorations. It is to try to elucidate what is novel about them and to connect them to broader debates and issues. What gene drives offer is a new form of "conservation," one that focuses on remaking the target of conservation rather than (or in addition to) addressing the social and ecological conditions around it. This is why gene drives are qualitatively different from other forms of conservation genetics, such as conservation cloning or synthesizing genomes to diversify populations, as well as from other types of interventionist conservation practices, such as assisted colonization. This qualitative difference warrants attentiveness and carefulness. As discussed above, the inference from "history is not a guide" to "we should creatively engineer systems" is too hasty, just as is the inference from "there is enormous anthropogenic change" to "embrace the Anthropocene."

To see the potential significance of this new conservation perspective, consider the genomic turn in medicine. Genetic medical technologies are powerful tools.

But they have also led to a substantial restructuring of medical practice—e.g., the skills, knowledge systems, activities, equipment, infrastructure, costs, approaches, and experiences of practitioner, patient, and backroom professionals. Moreover, these medical technologies have led to a partial reconceptualization of health and illness by normalizing and pathologizing genetic characteristics. Health and illness are often now conceptualized at the genomic level, as characteristics of a person's genome.

A genetic turn in conservation—and gene drives in particular—may produce something similar. Instead of looking at the conditions in which species populations are found, interventions could focus on the biology of the organisms themselves. This is a whole new way of conceptualizing the problem. It is not only that habitat has been degraded, but that the biology of the organism is not attuned (adapted) to its ecological environment. Hence, it is not only the modes of intervention, skills, knowledge base, and techniques that are novel, but the way in which conservation problems are approached. One concern is that this shift in perspective would lead to a sort of pathologizing of threatened species, implying that the problem is that they are not well fitted to the world. But, of course, the problem is that we have made a world that is not hospitable for them. Corals, bats, and amphibians do not have defective genomes.

In general, conservation genomics and synthetic biology pull toward a gene-oriented analysis of, and approach to, environmental problems. But gene drives are the true enabler of this kind of approach, since they allow engineered solutions to reach into the wild organisms themselves. This allows humans to adapt them to us and to our anthropogenic world, rather than requiring us to adapt our lifestyles and production systems to accommodate them. It is evolution by artificial selection among engineered variations: a full embrace of the Anthropocene.

For those who consider the ultimate cause of ecological destruction to be human-too-muchness and too-manyness, and the ideological cause to be the view that the ecological world should bend to the needs and wants of people, this will be disturbing. According to this view, gene drives are more of the same human arrogance, hubris, and domination that generated our conservation problems in the first place. It is an acceleration of the trend toward human-manufactured "nature." Moreover, if we are no longer concerned about conserving the natural world as we found it, but in making it as we think it needs to be, then what is the purpose of conservation? What, exactly, is it that we are trying to conserve? And what is the value of doing so? Certainly not naturalness. But the value of biodiversity also depends, in part, on the source of diversity being human-independent. If biodiversity is valuable in itself, then we ought to continually translocate species (whether or not they are climate-threatened) and engineer novel species in order to increase biodiversity and thereby the value of ecosystems (so long as the species are not disruptive); and we should try to make species-poor areas, such as deserts, richer in biodiversity. Even gene drive proponents would not assent to these, which suggests that the value of biodiversity is contingent on its source. Evolution of wild populations by artificial selection among engineered variations would be an amazing techno-scientific accomplishment, but it might not conserve the values that have always been associated with biological conservation: human-independent processes and naturally evolved diversity.

Conclusion

In this chapter, I have analyzed the ethical dimensions of using gene drives in the field for species conservation purposes. I began by arguing that the fact that we are in the Anthropocene (if it is indeed a fact) does not imply that people should embrace the role of intensively designing and managing the ecological world. Instead, it is consistent with the Anthropocene that we take a more restraintful approach. Therefore, the Anthropocene alone does not license deployment of gene drives.

I then suggested that there are two important perspectives from which to ethically analyze gene drives in conservation. The first is as a novel conservation tool, one that enables conservation biologist to effectively accomplish familiar goals, such as invasive species eradication, within a familiar conservation framework. When considered in this way, as a local conservation intervention, there are likely to be cases where gene drives are evaluated favorably on familiar decision-making criteria in comparison to alternative conservation approaches and techniques. However, these cases are not the conservation dilemma cases associated with macro-scale ecological change.

The second perspective is to see gene drives as enabling a novel form of conservation, one that allows conservation biologists to intervene into the biology of the organisms of at-risk species, rather than (or in addition to) the familiar practice of managing these species' ecological and social context. This type of application does not score as favorably on traditional conservation criteria, but does have the potential to address the conservation dilemma cases. When considering using gene drives in this more creative way, we must be attentive to how this technological power, particularly in combination with other tools in conservation genetics, could restructure aspects of conservation practice, including the point and value of conservation.

Analysis is not evaluation. The primary goal of this chapter is to begin to understand the ethical dimensions involved with a potentially powerful new technology—gene drives—so that informed assessments can be made. It is not to defend a position on what ought to be done with respect to gene drives, either in general or in particular cases.

Finally, it is important to note that while emerging genetic tools for conservation can be powerful, we should not overestimate them. Even with such technologies, there are no good adaptation options that can begin to scale to the extinction crisis that we are beginning to face. From the perspective of the value of species and biodiversity, as well as global and intergenerational justice, the best conservation strategy, by far, remains reducing the number of species that are at risk; and the only way to do that is to limit the scale of climatic and ecological anthropogenic change through technological innovation and modifying our behaviors, practices, and systems (Sandler 2012a, 2012b). Gene drives are not a silver bullet; they must also not become a moral hazard.

Notes

1 The material in this section is adapted from Sandler (2012a, 2013a).
2 The material in this section is adapted from Sandler (2018).

References

Ackerman, Diane. 2014. *The Human Age: The World Shaped by Us*. New York: W.W. Norton.

Atkinson, Ian A.E. 1985. "The Spread of Commensal Species of Rattus to Oceanic Islands and Their Effects on Island Avifaunas." In *Conservation of Island Birds. International Council for Bird Preservation Technical Publication No. 3*. Edited by Philip J. Moors, 35–81. Cambridge, UK: International Council for Bird Preservation.

Baillie, Jonathan, Craig Hilton-Taylor, and Simon N. Stuart. 2004. *2004 IUCN Red List of Threatened Species: A Global Species Assessment*. Cambridge, UK: International Union for Conservation of Nature. Available at: http://data.iucn.org/dbtw-wpd/html/Red%20List% 202004/completed/cover.html

Camacho, Alejandro. E. et al. 2010. "Reassessing Conservation Goals in a Changing Climate." *Issues in Science Technology* 26: 21–26.

Campbell, Karl J. et al. 2015. "The Next Generation of Rodent Eradications: Innovative Technologies and Tools to Improve Species Specificity and Increase their Feasibility on Islands." *Biological Conservation* 185: 47–58.

Ceballos, Gerardo et al. 2015. "Accelerated Modern Human-induced Species Losses: Entering the Sixth Mass Extinction." *Science Advances* 1: e1400253.

De Vos, Jurriaan M. et al. 2015. "Estimating the Normal Background Rate of Species Extinction." *Conservation Biology* 29: 452–462.

Doherty, Tim S. et al. 2016. "Invasive Predators and Global Biodiversity Loss." *Proceedings of the National Academy of Sciences* 113: 11261–11265.

Donlan, C. Josh et al. 2006. "Pleistocene Rewilding: An Optimistic Agenda for Twenty-First Century Conservation." *The American Naturalist* 168: 660–681.

Ellis, Erle. 2009. "Stop Trying to Save the Planet." *Wired*. May 6. Available at: https:// www.wired.com/2009/05/ftf-ellis-1

FAO. 2014. *The State of the World Fisheries and Aquaculture 2014*. Rome, Italy: Food and Agriculture Organization.

FAOSTAT. 2015. *Food and Agriculture Organization of the United Nations Statistics Division*. Available at: http://faostat.fao.org

Foden, Wendy B. et al. 2013. "Identifying the World's Most Climate Change Vulnerable Species: A Systematic Trait-Based Assessment of all Birds, Amphibians and Corals." *PLoS ONE* 8: e65427.

Friends of the Earth. 2016. "Genetic 'Extinction' Technology Rejected International Group of Scientists, Conservationists and Environmental Advocates." September 1.

Available at: www.foe.org/news/news-releases/2016-08-genetic-extinction-technology-rejected-by-international-group-of-scientists

Galloway, James N. et al. 2004. "Nitrogen Cycles: Past, Present, and Future." *Biogeochemistry* 70: 153–226.

Harris, James. A., Richard J. Hobbs, Eric Higgs, and James Aronson. 2006. "Ecological Restoration and Global Climate Change." *Restoration Ecology* 17: 170–176.

Higgs, Eric et al. 2014. "The Changing Role of History in Restoration Ecology." *Frontiers in Ecology and the Environment* 12: 499–506.

Hobbs, Richard J., Eric Higgs and James A. Harris. 2009. "Novel Ecosystems: Implications for Conservation and Restoration." *Trends in Ecology and Evolution* 24: 599–605.

Hoegh-Guldberg, Ove et al. 2008. "Assisted Colonization and Rabid Climate Change." *Science* 321: 345–346.

Krausmann, Fridolin et al. 2013. "Global Human Appropriation of Net Primary Production Doubled in the 20th Century." *Proceedings of the National Academy of Sciences* 110: 10324–10329.

Marris, Emma. 2013. *Rambunctious Garden: Saving Nature in a Post-Wild World.* New York: Bloomsbury.

Minteer, Ben A. and James P. Collins. 2010. "Move it or Lose it? The Ecological Ethics of Relocating Species under Climate Change." *Ecological Applications* 20: 1801–1804.

National Academies of Sciences, Engineering, and Medicine. 2016. *Gene Drives on the Horizon: Advancing Science, Navigating Uncertainty, and Aligning Research with Public Values.* Washington, DC: The National Academies Press.

NOAA. 2015. *What is Ocean Acidification?* National Oceanic and Atmospheric Administration. Available at: www.pmel.noaa.gov/co2/story/What+is+Ocean+Acidification%3F

Ricketts, Taylor H. et al. 2005. "Pinpointing and Preventing Imminent Extinctions." *Proceedings of the National Academy of Sciences* 102: 18497–18501.

Russell, James C. and Nick D. Holmes. 2015. "Tropical Island Conservation: Rat Eradication for Species Recovery." *Biological Conservation* 185: 1–7.

Sandler, Ronald. 2010. "The Value of Species and the Ethical Foundations of Assisted Colonization." *Conservation Biology* 24: 424–431.

———. 2012a. *The Ethics of Species.* Cambridge, UK: Cambridge University Press.

———. 2012b. Global Warming and Virtues of Ecological Restoration. In *Ethical Adaptation to Climate Change.* Edited by A. Thompson and J. Bendik-Keymer, 63–79. Cambridge, MA: MIT Press.

———. 2013a. "Climate Change and Ecosystem Management." *Ethics, Policy, and Environment* 16: 1–15.

———. 2013b. "The Ethics of Reviving Long Extinct Species." *Conservation Biology* 28: 354–360.

———. 2018. *Environmental Ethics: Theory in Practice.* New York: Oxford University Press.

Syvitski, James P. and Albert Kettner. 2011. "Sediment Flux and the Anthropocene." *Philosophical Transactions of the Royal Society A: Mathematical, Physical and Engineering Sciences* 369: 957–975.

Tripati, Aradhna K., Christopher D. Roberts, and Robert A. Eagle. 2009. "Coupling of CO_2 and Ice Sheet Stability over Major Climate Transitions of the Last 20 Million Years." *Science* 326: 1394–1397.

UNEP-WCMC and IUCN. 2016. *Protected Planet Report 2016.* Cambridge, UK: UNEP-WCMC. Available at: https://wdpa.s3.amazonaws.com/Protected_Planet_Reports/2445%20Global%20Protected%20Planet%202016_WEB.pdf

United States Climate Change Science Program. 2008. *Preliminary Review of Adaptation Options for Climate-Sensitive Ecosystems and Resources*. Washington, DC: Environmental Protection Agency. Available at: https://cfpub.epa.gov/ncea/risk/recordisplay.cfm?deid=1 80143&CFID=79321902&CFTOKEN=28197978

Vucetich, John A., Michael Paul Nelson, and Chelsea K. Batavia. 2015. "The Anthropocene: Disturbing Name, Limited Insight." In *After Preservation: Saving American Nature in the Age of Humans*. Edited by Ben A. Minteer and Stephen J. Pyne, 66–73. Chicago, IL: University of Chicago Press.

Wilkinson, Bruce H. and Brandon J. McElroy. 2007. "The Impact of Humans on Continental Erosion and Sedimentation." *Geological Society of America Bulletin* 119: 40–156.

WWF. 2014. *Living Planet Report 2014*. Gland, Switzerland: World Wide Fund for Nature.

Chapter 3

Gene Drives, Nature, Governance

An Ethnographic Perspective

Irus Braverman

> We are, at long last, learning to speak the language of nature. With sufficient cooperation and humility, we might even use it wisely.
>
> Smidler, Min, and Esvelt 2016, n.p.

Although evolution has enabled some naturally occurring genes to propagate above their expected frequencies, the recent discovery of CRISPR-Cas9 has enabled geneticists to cause this to happen at exceptionally high rates for chosen genes in the form of "gene drives." By encoding the CRISPR editing system into the DNA of certain organisms, geneticists can make a desired edit reoccur in each generation, driving the trait through the entire population by modifying only a few individuals (Esvelt et al. 2014; Harmon 2016).

Synthetic gene drives hold incredible promise, according to certain scientists at least. It is said that this technology has the potential to wipe out infectious disease vectors such as malaria-bearing mosquitoes, which kill hundreds of thousands of African children and adults each year, and thus to massively impact human health, especially in developing countries. Some have also suggested that gene drives could assist in the conservation of endangered species by eliminating the parasites and invasive predators which threaten their existence, such as in the cases of sylvatic plague for black-footed ferrets (Novak 2016) and of invasive rodents on islands (National Academies of Sciences, Engineering, and Medicine 2016). Gene drives could, to take another example, suppress the avian malaria-bearing mosquitoes which transfer disease to endangered honeycreepers in the Hawaiian Islands (ibid.). By using targeted modifications within an organism's DNA in order to impact certain populations or even entire species, gene drives could minimize the ecological effects of insecticide and rodenticide use—thus providing what some say are clear advantages for the natural environment.

Alongside what is presented by scientists and the media as an incredible promise, the introduction into the wild of individuals containing gene drives involves major risks. "DDT only goes where you spray it, [but] a gene drive will keep going beyond; unless you design it otherwise," geneticist Andie Smidler told me in an interview at the Harvard School of Public Health (Smidler 2016). Along these lines, a report issued by the expert committee convened by the National Academies of Science, Engineering, and Medicine in June 2016 warns about the "cascade of population dynamics and evolutionary processes" that might be initiated by the introduction of an artificial gene drive into the wild (National Academies of Sciences, Engineering, and Medicine 2016). Even with sophisticated computer models and ecological risk assessments, such a cascade of effects would be difficult to predict, according to the report.

Given the intentional design of gene drives to spread like fire and that of impacted organisms to cross jurisdictional borders, the National Academies of Sciences report cautions that there is currently no effective regulatory mechanism to oversee gene drive research and field trials (ibid.; see also Akbari et al. 2016; Charo and Greely 2015; Oye 2015). As a result, some have pointed out that in the context of genetic engineering, "the implementation of social responsibility in the United States has been left virtually solely to scientists" (Sankar and Cho 2015, 23). It is in this regulatory void that the operating scientist becomes a self-regulator, her values and visions that much more important as they at least partly determine the scope of the research that she will undertake and its normative dimensions. A new scientist-regulator hybrid thus emerges at the center of gene technology governance.

My chapter moves away from the usual discussion of gene drives to consider the emotional and relational landscape of gene drive scientists. Specifically, I draw on in-depth discussions with several gene drive scientists and related experts to explore their personal perceptions of nature and their relationship toward the animals they edit and engineer. Arguably, the underlying relationship of the scientists I have interviewed with nonhumans and the environment informs, even paves, their research path. Their perceptions about nature in particular provide these scientists with the normative justification to pursue research that would alter not only individual DNA but also entire populations and the ecosystem. This chapter seeks to discover how gene drive scientists think and feel about the impact of their research.

I will begin this ethnographic contemplation with a hands-on description of how gene drives work and a brief account of the ecological implications of this technology. Next, I will discuss the attitude expressed by prominent gene drive engineer Kevin Esvelt toward nature and about the relationship between humans and nonhumans. Finally, I will present my conversations with three biologists who are working on gene drives in mosquitoes about the role of nature and the value of life. I should note at the outset that this is by no means an exhaustive study of gene drive scientists, nor is it necessarily representative of the larger gene editing community. Nonetheless, the discussions below expose some of the major emotional and relational questions and concerns facing contemporary scientists in the nascent field of gene editing and how they conceive of and implement the self-regulation of their research.

The Labor of Gene Drives

The actual work of gene drive scientists didn't hit home until I sat with Andie Smidler. Smidler is a Harvard School of Medicine graduate student working with two prominent scientists, Kevin Esvelt and Flaminia Catteruccia, and an unusually energetic woman. She doesn't have an office, which is why we convene in the kitchen behind Catteruccia's lab at the Harvard School of Public Health. When I complain to Smidler about how difficult it has been to wrap my head around gene drives, she readily takes on the challenge. "Everybody has two copies of every gene, one from mom and one from dad," she tells me matter-of-factly. "These genes are usually very similar, but where they differ is where you get all different types of humans. Dad has a gene for tallness and mom has a gene for shortness, for example." I am able to follow her to this point, but then things get more complicated:

> A cell normally repairs its DNA by searching for the other copy to say, "Hey, I'm broken, I need to fix myself, but that one is still intact, I bet that's the right gene." So [if mom's gene is broken,] dad's gene drive cuts mom's empty DNA and then mom's broken DNA is like, "I need to fix myself." So the cell has a natural DNA repair mechanism. And it uses DNA from the [synthetic] gene drive as the template for repair. And it says, "I'm replicating to fix myself, but oh hey look, this gene drive looks similar enough, I'll copy it over," and it copies it over. So you're born with one copy, and now you have two copies. If this happens in the germline . . . what you get is biased inheritance. Now that both mom and dad's copies are gene drives, all of their kids are going to get one copy of the gene drive, and then it does the same thing in the kid as it did in the parents. It cuts the other version, copies itself over and it keeps spreading that way. So it artificially gets inherited at a higher frequency than normal. And that's the basic concept behind most of these modern gene drives.
>
> (Smidler 2016)

The process of designing a stable drive is apparently a more complicated matter altogether. Smidler explains:

> To make an evolutionarily stable drive, you have to put the drive into these really critically sensitive genes in a very specific way. And you have to engineer them so precisely because you have to insert this massive genetic cargo in the middle of this gene that normally can't be disrupted. You design it so that it can survive, but it's very tedious and difficult to do. These are genes that you can't cut in the first place and have the cell survive, but to build our gene drive, not only do we have to cut it, but then we have to do all this crazy genetic engineering stuff, get the cell to live, and then get the product that we want. So it took us the last two and a half years to be able to make the genetic engineering change in that essential gene and get that cell to live. We

have it now. We don't have the drive. To build the drive, we have something called a docking line, and basically it's a sequence that you can put on your transgenes.

(ibid.)

This account affords a glimpse into the on-the-ground trials of gene drive scientists. It is surely the closest I've come to comprehending the scientific labor involved in gene drives on the micro-scale of the DNA and the immense battles over harnessing life into a digitized and predictable form (see also Chapter 9, this volume).

The Ecologies of Gene Drives

Beyond the individual, gene drives could also impact life at the level of populations, species, and ecosystems. Science and technology studies (STS) scholars Javier Lezaun and Natalie Porter refer to gene drives as a "transgenic technology." They explain that "transgenic technologies hope to find in the genome of the pertinent animal species a molecular 'switch' that would short-circuit transmission to humans" (2015, 4). Contrasting this with the One World, One Health (OWOH) approach, which attempts to contain the circulation of pathogens among species, Lezaun and Porter caution that genetic modification "promises a kind of directed animal evolution, which would absolve humans from the need to alter their behavior in the service of disease prevention" (ibid., 97). Instead of alertness to the animal–human interrelations prescribed by the OWOH model, transgenic technologies present "entirely new dynamics of intra-species competition," giving rise to "qualitatively different human-animal ecologies" (ibid., 99; see also Chapter 2, this volume). Along these lines, in his in-depth documentation of the path of dengue fever, mosquitoes, health providers, and poor communities in Nicaragua, STS scholar Alex Nading develops the term "the politics of entanglement." This term serves Nading to highlight the strife of local communities "to remain alive to the world around them despite global health strategies that seek to insulate them from their environments" (2014, back cover).

The National Academies of Sciences, Engineering, and Medicine expert committee on gene drives seemed less apprehensive of transgenic technologies generally and of the disentanglements that may occur as a result of their anthropocentric focus in particular. After a year-long study of six potential sites for gene drive use, the long-awaited report, released in June 2016, called for "carefully controlled field trials," effectively giving experiments outside the lab a green light to proceed. The report nonetheless cautions that:

before field testing or environmental release of gene-drive organisms, it is crucial to establish a detailed understanding of the target organism, its relationship with its environment, and potential unintended consequences. It is also essential to consider confinement and containment strategies to reduce the potential for unintended releases.

> (National Academies of Sciences, Engineering, and Medicine 2016, 2)

Harvard biologist Flaminia Catteruccia, who is part of the small team developing gene drives for nonhuman species, responded to the report:

> If you use a gene that kills the plasmodium parasite in mosquitoes, how will it behave with other pathogens? Will it affect, for instance, insecticide resistance? If you affect dengue, how will that behave with yellow fever or, now, with Zika virus? . . . There are lots of questions to be addressed before we can safely release them.
>
> (Quoted in Powell 2016)

One concern is that the drive could "jump" to other organisms besides the targeted ones (Ledford and Callaway 2015).

Importantly for this chapter's purpose, the National Academies of Sciences report also briefly considered the broader role of nature and the meaning of human-nonhuman relationships when it adopted the following statement on human values:

> Perspectives on the place of human beings in ecosystems and their larger relationship to nature—and their impact on and manipulation of ecosystems—have an important role in the emerging debate about gene drives. The increased power for human beings to alter wild species and perhaps to eliminate them, thereby altering the shared environment—will be intrinsically objectionable to some people. Proposals to use gene drives in ways that might lead to the extinction of species or significantly alter the environment will require especially careful review.
>
> (National Academies of Sciences 2016, 18)

Only a couple of months after the report was released, 30 environmental leaders who convened at the 2016 World Conservation Congress issued the following statement:

> Given the obvious dangers of irretrievably releasing genocidal genes into the natural world, and the moral implications of taking such action, we call for a halt to all proposals for the use of gene drive technologies, but especially in conservation.
>
> (Friends of the Earth 2016)

Founding signatories of this statement include Jane Goodall, David Suzuki, and Vandana Shiva.

I would propose that the differences in approach toward the desirability of gene drives and their regulation are caused by divergent foundational approaches toward the meaning and value of nature. Whereas environmental leaders seem to see an inherent and intrinsic value in the "natural world" and its conservation, and thus highlight the dangers in gene drive technologies (see Chapter 5), genetic engineers either do not see the value in conserving existing nature as is, or they do not see this world as natural to begin with. These bifurcated views about nature tend toward either cautionary or interventionist approaches, in turn engendering different modalities of governance.

Gene Drives versus Nature

Kevin Esvelt is a prominent scientist in the gene drive community and much respected within this community for his consistent call for responsible, responsive, open, and transparent science (see, e.g., Chapter 1, this volume). He is strongly influenced by utilitarian thinker Peter Singer and is quite outspoken about his views toward nature. Specifically, Esvelt is quick to assert the immorality of nature, which he sees as "red in tooth and claw," an expression he is particularly fond of. "Existence in the wild is basically unmitigated pain and suffering," Esvelt tells me as we sit in one of Boston's gardens, observing squirrels and rabbits.

> No ethics committee would ever approve the [creation of] organisms [who] live their lives in the wild, [are] eaten alive by parasites, [are] taken out by nasty diseases, and [are] constantly having to evade predators—with horrific suffering when the predator eventually catches you.
>
> (Esvelt 2016a)

Wilderness is immoral and evolution is immoral—or amoral—Esvelt seems to use these two terms interchangeably. He explains:

> Darwin probably lost his faith. He didn't quite say it, because it was politically impolite at the time, but he said he could not conceive of how a good creator could have created the [spider wasps], which are the wasps that paralyze their prey, and inject them with eggs that then grow into larvae and eat them from the inside—using no painkiller, of course. Why would you do that? That's not evolutionarily advantageous. So evolution . . . [is] outrageously cruel.
>
> (ibid.)

As part of his stance toward nature and evolution, Esvelt is most interested in directing his talents toward "urgently solving the problem of animal suffering and human suffering." "Once we get the ball rolling," he tells me, "then I'll be able to think about changing the cultural perceptions on suffering, and cultural perceptions of nature, and what is defined as pure and good." The romanticization of nature has always bothered him, Esvelt says in our interview. "People who have lived in cities long enough start to romanticize what is out there, because living in a city is bad, of course. Until very recently it was bad—it was disease-ridden, crime-ridden, life expectancy was shorter than in the countryside. [But] now it's flipped." Still, the grass stayed greener on the other side, according to Esvelt. And "that's where this notion of nature being the font of purity and goodness came from. Before that . . . the notion that you would go camping was insane. Why would you do that? Why would you give up everything that we've worked to achieve?"

Esvelt finally compares the suffering that occurs in nature to lemons, and human progress to lemonade. "I'm very much of the opinion that life is made of lemons and we're slowly making lemonade. I mean, more lemonade than we ever have before . . . We're better off now than we have been. I think that's pretty much indisputable." He then proceeds to ask rhetorically: "Would I rather exist in a world in which we tweaked things in order to reduce the nasty parasites that lay their eggs in you and those larvae eat you alive from the inside, or one in which we didn't?"

The notion that technology makes lemonade out of nature's lemons calls to mind John Locke's views on nature and progress that underlie modern theories of property. The idea that humans improve on nature also feeds into a regulatory approach that valorizes technological progress, with the only relevant concern being the proper allocation of the lemonade, and not the very existence of lemonade and the ever-growing need for more of it.

Alongside his views toward nature, another important aspect of Esvelt's philosophy that arguably impacts his work as a genetic engineer regards the relationship between human and nonhuman animals. Esvelt readily admits to being a speciesist in this context. In his words: "there are lots of different forms that [life] can evolve into, and some particular forms cause more negative net utility to other species than is counterbalanced by the positive utility of them existing" (ibid.).

Esvelt has been influenced by Brian Tomasik's philosophy, which contends that even if insects suffer 12 orders of magnitude less than humans, their suffering outweighs ours and every other mammalian life on earth (2015, 139). Nonetheless, because Tomasik believes that the insects' life lacks happiness and is outweighed by the pain of death, he eventually encourages the extinction of such insect populations.

While Esvelt does not agree with Tomasik's mode of counting insects versus humans (in his words: "I don't go that far, and that's partly because I grade things with cognitive complexity to a much greater extent"), he is clearly on board with the utilitarian mode of thinking that determines and distinguishes life's value based on degrees of suffering. When I am slow to follow his calculus of positive, negative, and neutral utility, Esvelt points to a baby rabbit that happened to cross the path behind our bench, stating:

> That baby bunny probably has it better off than most insects, depending on its current parasite load, and how it dies. It's probably better to be larger and comparatively cognitively sophisticated, in most cases, because you're likely to be longer lived and have a longer youth and positive utility.
>
> (Esvelt 2016a)

Using a similar calculus, Esvelt argues that because humans are the most sophisticated cognitive creatures on earth, our suffering and pleasures balance out in relation to those of insects, despite their incredible quantities. And although he admits that humans have been causing mass extinction to numerous life forms on this planet, he reasons that "if we terraform Mars and seed it with life, that will more than outweigh any of our past sins" (ibid.). While this may sound like a scene from a science-fiction movie, the project of populating Mars with life is already in progress at George Church's lab (Davis 2016), where Esvelt conducted his postdoctoral research.

Esvelt's utilitarian philosophy is not limited to living forms, however. "I try not to be too much of a materialist, or a life-ist," he tells me. "What's so special about life, anyway?" he asks, explaining that "in many ways, biological evolution is inferior to the cultural kind, and [to the] technological kind." There are other considerations to life and biodiversity, he continues. From his point of view:

> A virus is an interesting information string, but is it more valuable than an idea? An idea is a different form of evolutionary replicator; it's a different form of information. But why is the virus more or less valuable just because it happens to be encoded in DNA, rather than in neurons, or magnetic strips, or ink on paper? There are so many different potential forms of information, and what I'm interested in, literally, is informational patterns and complexity.
>
> (Esvelt 2016a)

And if life isn't all that special, death is definitely overrated, according to Esvelt. "I am one of those people who sees no particular purpose in death and I would abolish it if I could," he tells me in our interview. "Not because it's so tragic when you're dead, it's just unfortunate for everyone else and you can no longer gather positive experiences." Not only the quality of life, but also the very essence of life and its relationship to its other, death, are newly called into question.

But Esvelt was not always a utilitarian. "I was a radical environmentalist in high school," he admits. "I viewed humanity as extinguishing patterns that nature had created [and] that had their own intrinsic beauty." Generally, the intrinsic value approach to nature is juxtaposed with the instrumental value approach that sees nature as valuable only so far as it is useful for humans (Callicott 1995; see also Chapter 5, this volume). Esvelt tells me that his views had radically changed when he realized, in his early college years, that

> if what I really care about is the diverse patterns created by life, then humanity is the one species that's going to get life off this planet and create new forms of life. . . . [T]hat is more important than anything else; it's more important than conservation [on earth]. I've never seen anything particularly spectacular and

holy about the fact that most of the universe is barren [and] lifeless rocks, and more hydrogen. Life is far more interesting. Life evolves. Our overall purpose, insofar as we have one, is to spread interesting patterns throughout the universe. We have a duty to expand our garden, but we need to make sure that it is a garden, and not the gladiatorial arena [that wilderness is].

(Esvelt 2016a)

The scientist thus emerges as the knight whose responsibility it is to govern nature and improve on it. Furthermore, conservation is configured as a marginal concern in comparison to the much broader concern with life as a mode of information and its expansion beyond planet Earth.

The Nature of Transparency

Reading and re-reading the long transcripts of our conversations, I find it hard to reconcile Esvelt's almost-contemptuous attitude toward nature and all-encompassing utilitarianism with his call for full transparency in gene drive research and his insistence on democratic decision making in this regard. In his June 2016 contribution to *Nature*, Esvelt explains that "gene editing can drive science to openness." "Because the consequences of mistakes involving gene-drive organisms could affect communities outside the laboratory," he writes,

> scientists have an obligation to openly share their plans, invite suggestions and concerns, disclose experimental results as soon as possible, and redesign the technology as needed. Applied to gene drives, such an approach will also have a greater chance of earning popular support for applications that could save millions of human lives and rescue numerous species from extinction.
>
> (Esvelt 2016c)

In Chapter 1 of this volume, Esvelt repeated his concern that if scientists do not govern themselves properly by involving the relevant publics, such publics will lose their trust of these scientists and react by taking away their social and normative license to save the world.

Along the same lines, Esvelt is also concerned about the existing regulatory system and, in particular, about the way that patents and publications are managed, which disadvantages open science and punishes those who practice it. In his words:

> Sadly, open and responsive science flies in the face of current incentives. Scientists who disclose their ideas are often "rewarded" by being scooped by another lab, rather than by being recognized for their creativity. It is a prisoner's dilemma. The benefits come from cooperation by everyone. But by

participating, you risk being exploited by people who steal your idea, get it working before you do, and claim the credit.

(Esvelt 2016c, 153; see also Chapter 1)

James P. Collins is professor of biology at Arizona State University and the co-director of the expert committee that released the National Academies of Sciences, Engineering, and Medicine report on gene drives. Collins tells me in our interview that upon the report's release, Esvelt called him to complain that the transparency of research and the importance of biological containment were not emphasized enough, and that the public should always have a say about the introduction of gene drive organisms. Collins comments in response:

> When he says that residents need to have a say, Kevin's thinking [of] Lyme disease in Martha's Vineyard. But why stop at this island? Who is the relevant public? And what does "have a say" mean, anyway? Is it a vote? What kind of a vote? How will it all work?
>
> (Collins 2016)

When assigning a decision to a group of people, one must adhere to their decision, even if it is objectionable, Collins tells me, implying that he is not sure that Esvelt truly endorses this particular aspect of democratic decision making process. Esvelt responds with his own critique: "I personally believe the current system of do-everything-in-secret-until-submitting-the-product-to-regulators is a pretty terrible system for inspiring public confidence in the governance of technology" (Esvelt 2016b). While everyone seems to agree that the public must be involved, the question is, more precisely, who the public is—and who gets to decide on this question.

Another question that has until now remained at the margins of the discussion is whether the relevant "public" should include nonhumans. "At the Committee, no one really spoke for Nature," Collins admits.

> But what would that mean, anyway? I can't imagine how this would work. From the local frogs to the local birds to the mosquitoes themselves—they all have a stake. This is why it's an out-of-bounds question. Anyway, it didn't come up in our statement of task, and with the limited time we had, we needed to stay on task.
>
> (Collins 2016)

Over two decades ago, French STS scholar Bruno Latour called for a "parliament of things" that would take nonhumans into account in democratic decision making (Latour 1993). How to work around human agency and move toward nonhuman actancy has been a topic of much discussion among posthumanists (see, e.g., Kohn 2013) and one that could use a more serious discussion in this context, too.

Another curious aspect of Esvelt's philosophy of governance is his view on patenting gene drives. "I opened this dangerous box, therefore I better make sure it doesn't fall into the wrong hands," Esvelt says in response to my question about why he had applied for a patent on CRISPR gene drives (2016a). Yet despite his repeated calls for collaboration rather than competition and for openness in place of secrecy, during a gene drive conference in February 2016, Esvelt was reluctant to share the news about his application for a gene drive patent with the gene drive community. When I asked him about this, Esvelt blamed the patent system. In his words:

> The primary link between [intellectual property] and openness is that even though the patent system was explicitly set up to encourage disclosure, the current system makes it very hard to openly share your ideas without losing your ability to patent them, and therefore to give them away. We are forced to file provisionally and prophetically, then race to demonstrate that it works within a year, when we have to file the real thing.
>
> (Esvelt 2016b)

As I have mentioned in the introduction, Esvelt has sought to use patents to protect the public from what he conceives as dangerous private interests. Andie Smidler explains:

> Kevin [and I] decided to patent the CRISPR gene drive concept because . . . we wanted to . . . make sure that no one could try and turn it into a Monsanto situation, where everyone hates it, it's loathsome, it's terrible, it's mismanaged, [and certain] people are just itching for a profit.
>
> (Smidler 2016)

Here, not only does the hybrid role of the scientist-regulator emerge, but the responsible scientist also ends up representing the public interest in place of the incompetent state.

Gene Drives in Mosquitoes

To date, CRISPR gene drives have been used in four different species: yeast, fruit flies, and two different mosquito species (National Academies of Sciences, Engineering, and Medicine 2016, 1). Mosquitoes are by far the most urgently and comprehensively studied species for gene drive applications. Their genetic editing is also probably the most advanced experiment in gene drives. The goal of scientists has been to alleviate human diseases transmitted through mosquitoes, such as malaria, dengue, and Zika. The work of controlling certain mosquito species through genetic editing is not new: the for-profit company Oxitec has already paved the path for genetic work on mosquitoes through its introduction of sterile male mosquito populations. Oxitec has not been using gene drives, which means that they must perform periodical releases, which makes for a better business model than global gene drives (Alphey 2016).

Oxitec's "environmentally friendly OX513A male mosquitoes" were introduced in five separate efficacy trials—across Brazil, the Cayman Islands, Panama, and Malaysia—leading to a greater than 90 percent reduction in the local *Aedes aegypti* (disease-borne) populations (Oxitec 2016, 3). Once the "friendly male" mates with the relevant females, the offspring inherit a self-limiting gene and die before becoming functional adults, thereby reducing the size of the wild population. By 2016, over 150 million Oxitec mosquitoes were released in field suppression programs. A built-in fluorescent marker was designed into the engineered mosquitoes in order to track them in the field and the lab (see, e.g., Figure 2).

Although the local community in South Florida refused Oxitec's engineered mosquitoes in the recent past, a new debate is currently unfolding about the release of a new type of Oxitec mosquito to Key Haven in order to control the spread of Zika. In August 2016, the FDA found "that the proposed field trial will not have significant impacts on the environment" (FDA 2016). Being the hot potato that it is, the mosquito control board nonetheless decided to defer the final decision to "the people." A referendum held in fall 2016 produced split results: while the voters of Monroe County voted 58 to 42 percent in favor of releasing the genetically modified mosquitoes, the community of Key Haven, where the mosquitoes would be released, opposed the resolution by a 65–35 margin (Langston 2016).

Flaminia Catteruccia is a molecular entomologist who studies mosquito reproduction and conducts research on gene drives at the Harvard School of Public Health. She has been part of the Harvard collaboration on gene drives in mosquitoes that also includes Esvelt, Smidler, and Church. We spoke at her office, and then she showed me her lab, abuzz with thousands of mosquitoes. As with Oxitec's approach, rather than kill the baby mosquitoes, Catteruccia's model is to null the reproductive act—namely, to have sterile males mate with females. Since the females mate only once in their lifetime, such females will not have offspring.

Catteruccia's program does not kill any mosquitoes, only limits their reproduction. She is proud of her nonlethal method, admitting to having a special relationship with the mosquitoes she works with. "We originally captured those mosquitoes and then we colonized them and brought them back to the lab," she tells me about the origin of her lab's mosquitoes. "So these are [lab] colony mosquitoes," she continues. "But from time to time, you might want to refresh your colony with a few mosquitoes [from the wild] to make it as field-like as possible" (Catteruccia 2016a).

"Mosquitoes are the least-liked animals in the world," Catteruccia laments. "People are disgusted by rats, but I think everyone hates mosquitoes." As for her own feelings toward mosquitoes, while she admits that the first time she saw a mosquito under a microscope was scary ("It had big eyes, and was furry, it was a bit like, 'What am I looking at?'"), Catteruccia has since then fallen in love with this organism. "There are 5,000 different mosquito species, but only a handful of them transmit malaria," she says in their defense. "We study them because they are a fascinating organism that transmits a disease that kills. They don't kill, themselves, they just carry these pathogens," she continues. "The whole world would be a very sad place without mosquitoes," she adds. And while "there are not many organisms that feed on a specific and exclusive mosquito diet," eliminating mosquitoes would definitely change the environment, according to Catteruccia. Birds both feed on mosquitoes and determine their migration patterns partly in order to avoid their bites. An elimination of mosquitoes would affect the food chain in the other direction, too, as mosquitoes feed on bacteria, fungi, and algae. But there isn't much concern that mosquitoes will become extinct, Catteruccia assures me. In her words: "we're never going to get rid of mosquitoes. They're going to outlive us. They were on this planet 200 million years ago" (ibid.; see also Webber, Raghu, and Edwards 2016).

Catteruccia studies the interactions between male and female mosquitoes, among the females, and between the mosquitoes and the disease-bearing parasite. "So it's a really a complex series of interactions that you study *in vivo*, in the whole organism. And that's what I love about this," she reflects. In her lab, Catteruccia houses thousands of mosquitoes at one time, from which she takes 50 to 100 females and infects them with the malaria parasite to conduct genetic experiments. She explains the multiple layers of physical containment for the infected mosquitoes: "There's a room within a different room, [and] within this room there is a glovebox. And then the [infected mosquitoes] are in cages within the glovebox. So there are four levels of separation from the rest of the population." When she or her staff work on the mosquitoes, they always use gloves, and they always handle the mosquitoes within the cages. "The only way they come out of the cage is when they are dead," Catteruccia assures me. The geographical design of the lab embodies normative codes of confinement and separation (ibid.; for more about physical and biological containment, see Chapter 4, this volume).

In light of her high appreciation of this animal, how does she feel about killing mosquitoes? I ask. Catteruccia responds:

> I actually had an undergrad student who was doing a practical on mosquitoes. He refused to kill them. And that was the first time that [this] happened to me. Personally, there are some things that I don't like doing to them. And I always apologize when I do it. So I say, "Sorry," and [then] I chop their heads off, for instance. It's true! It's ridiculous, and I never thought about it. But now that you ask me, there is an element that is not 100 percent neutral for me. There is always "I'm killing a living organism," there is a little bit of that. But, they don't live very long, so there's that element—I killed them [just] a week before they would die anyway. And also, we keep them in captivity and these ones definitely wouldn't [exist without our breeding]. The way we kill them is very fast. Not so much to alleviate suffering, but to make sure they don't fly around. So something that kills them very fast is to decapitate them. With needles. We anesthetize them with carbon dioxide, and then they go to

sleep. And then you can kill them. I don't think anyone enjoys it. . . . We do it because we have to. We never discussed this, maybe we should?

(ibid.)

Catteruccia goes on to compare lab work on mosquitoes and mice. She never wanted to work with mice, she admits, mainly because she couldn't deal with killing them. Catteruccia reflects about the regulatory aspect of working with these two animals in the lab:

The research on mice is much more regulated than with mosquitoes. No one asks you to kill mosquitoes humanely. For mice, you have to have certain protocols. I really think it goes hand-in-hand with our knowledge of their nervous system. Mosquitoes are much less sophisticated, so their suffering is more limited.

(ibid.)

Since they suffer less, it is implied, the mosquitoes' life is worth less—yet another aspect of the utilitarian approach so prominent in Esvelt's philosophy and apparently underlying many of his fellow biologists' way of thinking. Another prominent gene drive biologist tells me along these lines: "A lot of biologists have to kill their subject matters; it's something we're accustomed to" (James 2016). Underlying these statements by scientists whose daily research involves nonhuman animals is the assumption that humans are superior to these animals. A second assumption is that humans must kill (certain animals) in order to make (certain humans) live. Third, animals are ranked on a scale that renders some more equal than others. The death of certain species is grievable, while others are merely killable (Braverman 2016). This biopolitical construction of relative worthiness is supported by existing regulations and guidelines, the value ascribed to certain animals translated into their mode of governance, and vice versa.

Esvelt fills me in on another aspect of the mundane genetic lab work with mosquitoes. He explains that because mosquitoes breed in swarms, getting two particular mosquitoes to mate is a problem. The way that mosquito scientists get around this is to

take the male, lop his head off, then mount him, and [finally] knock out the female and forcibly mate her to the male. So behead the males, use drugs to knock the females unconscious, and forcibly mate the unconscious female to the dead headless male.

(Esvelt 2016a)

Esvelt highlights the necessity of this procedure:

A couple of times, the mosquito that [we] wanted was a rare male. So out of all these injections, only one male had the right fluorescent signature.

You pull its leg off to sequence it, and it's like, "Yes, he's the one, we just need him to mate." And so you do the mating and you hope that she would find him, and he's on his last legs and clearly dying, and [we're] like, "Welp. Okay, I'll pull off the head and find an unconscious female." And that works. So yes, necrophilic rape is a thing that we do to mosquitoes. . . . To someone whose moral foundations are primarily based around harm and care—for example, a consequentialist—there is absolutely nothing wrong with this [practice]. The female mosquito is unconscious, will never know what happened, and being equipped with only the nervous system and cognitive capacity of a mosquito, certainly isn't programmed to react negatively in any way, even if shown a video of what occurred. Indeed, the very notion that a mosquito could be traumatized in that way is to anthropomorphize to a rather remarkable—and wholly inaccurate—extent.

(ibid.)

Catteruccia strongly disagrees with Esvelt's description of mosquito lab work and is particularly rejecting of the term "necrophilic rape." "I found the term inaccurate and sensationalist," she writes me in an e-mail, explaining that

some mosquitoes don't mate in captivity and the only way to maintain a colony and study them is to induce mating artificially, which is done by removing the male's head and putting the female to sleep. The male is still physiologically alive when mating, so the term is at least inaccurate. The female is not stressed by the procedure, and behaves the same, lives as long, and reproduces as well as freely mating females.

(Catteruccia 2016b)

It is important to note that although her recent career has been focused on conducting genetic research on mosquitoes, Catteruccia nonetheless acknowledges that genetically tinkering with this organism to address human health issues is not the best solution from a societal point of view. Instead, she admits, it would have been much better to attend to conditions of poverty, inequality, and lack of education that have created such pockets of human vulnerability to mosquitoes in developing countries in the first place. But that seems to be a much harder challenge to address than fixing genes in mosquitoes, Catteruccia laments in our interview (Catteruccia 2016a).

Another molecular entomologist, Anthony James, has been working with gene drives and malaria for quite some time. He first became interested in this work when, as an undeclared undergraduate, he happened to get a job washing dishes at the local drosophila lab. Although he talks about collaborations within the small community of molecular entomologist working on mosquito-borne disease (less than a dozen around the globe, he points out), James is nonetheless convinced that his lab's approach of altering mosquito populations makes much more sense than Catteruccia's approach of suppressing them. "With the

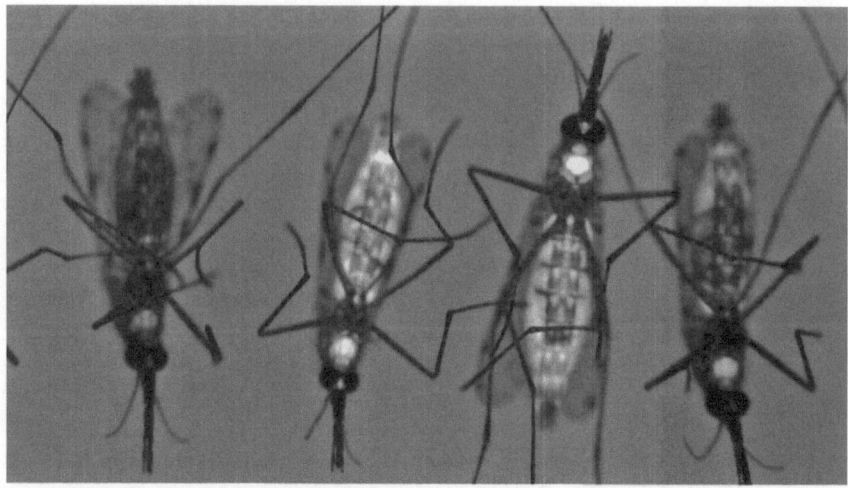

Figure 2 Adult female mosquitoes (*Anopheles stephensi*), vectors of urban malaria in India, transformed as part of a background experiment for gene drive.

Courtesy of Anthony James.

suppression approach they will always come back, so this solution is not sustainable," he explains in our interview.

James also emphasizes that his use of the term sustainability does not carry any ecological meaning. "If I could, I would kill them all, but it is just impractical," he clarifies. Although he doesn't see mosquitoes as charismatic animals, James admits that they are exquisitely beautiful. Still, "would I want to mess with a beautiful tiger that's coming to attack me?" he asks, emphasizing that in his view, what we're dealing with here are dangerous animals. "When a female mosquito is fixated on you, it's either you or her," he tells me. As for how he feels about tinkering with nature, James admits that he grew up with the "back to nature stuff," so he is "quite respectful toward nature." At the same time, since he doesn't perceive the mosquitoes' environment as natural, but rather as human-created, he is not very concerned about tinkering with it. And as for his feelings about our rights, as humans, to engineer the molecular basis of life, James responds by referring me to the same day's headlines about the recent epidemic of lethal yellow fever. "We must respond," he tells me, implying that tinkering by scientists is not only a right but also a duty.

Conclusion

Synthetic gene drives raise ethical, ecological, and legal questions that are so broad and consequential that they can be difficult to grasp. What is clear, however, is that the power to directly alter not just a singular form of life but also the genetics of entire populations and species are currently both under-regulated and under-theorized. In place of state regulations, what seems to be emerging is a form of self-regulation by the gene drive scientists themselves. My chapter has drawn on in-depth interviews with several prominent gene drive scientists to explore their approach toward nature, animals, and the environment. My assumption has been that their approach impacts and regulates the way they work, and this assumption has been confirmed through the more personal stories that each of these scientists has generously shared with me. Although they have not contemplated these issues to the same degree, a few common assumptions about the role of nature and about animal-human relations did emerge from the interviews—most prominently, the notion that killing insect populations, modifying their genes, and impacting the planet's ecological systems are justified in order to reduce human suffering and produce novel ecosystems. This is by no means a revelation; it is consistent with how humans have been treating insects, and nature more broadly, for centuries (see Chapter 2, this volume).

Nonetheless, the nuances are what have made this story interesting, and the differences between the scientists I have interviewed have revealed some of these nuances. What makes more sense: to sterilize mosquitoes or to alter their genes? How do scientists feel about, and go about, such killings, and how do they explain it to themselves? And finally: how do gene drive scientists view and regulate their power to genetically impact entire populations and ecosystems? I was struck by the differences I encountered between the biologists and the genetic engineers among my interviewees. Although my "n," or sample size, is clearly nothing close to statistical, my sense from the in-depth interviews was that as a biologist, Catteruccia holds a more intimate relationship with her animal "subjects," considers their suffering, and is sensitive to the possible conservation impacts of their modification. On the other hand, as a synthetic biologist and genetic engineer, Esvelt's thinking is more abstract and utilitarian. He is willing, happy even, to tinker further. Toward the end of our interview, Catteruccia encapsulated these differences:

> These hardcore geneticists are so provocative, in a way, [that] my innate instinct is, "Whoa, Whoa!" . . . I'm a more of a molecular entomologist. So I study insects, I'm definitely not hardcore like that. These guys are amazing with what they come up with and what they can do, with what they develop. With me, I'm just using those tools to study basic biology.
>
> (Catteruccia 2016a)

A final thought: while the gene drive scientists I've interviewed were all highly intelligent, creative, and well-meaning, they were at the same time clearly under-educated in all matters ecology related. Their views about nature-human-animal relationships could benefit from some sophistication and historical contextualization. Maybe a workshop about nature with prominent cultural thinkers and ecologists would be a practical policy recommendation for this group of people, who in many respects hold the future of gene drive technology and its governance in their hands.

References

Akbari, Omar S. et al. 2016. "Safeguarding Gene Drive Experiments in the Laboratory." *Science* 349: 927–929.

Alphey, Luke. 2016. Founder, Oxitec. Personal communication, Raleigh, NC, February 23.

Braverman, Irus. 2016. "Anticipating Endangerment: The Biopolitics of Threatened Species Lists." *BioSocieties*. DOI: 10.1057/s41292-016-0025-0

Callicott, J. Baird. 1995. "Intrinsic Value in Nature: A Metaethical Analysis." *Electronic Journal of Analytical Philosophy* 3.

Catteruccia, Flaminia. 2016a. Associate Professor of Immunology and Infectious Diseases, Harvard University T.H. Chan School of Public Health. On-site, Boston, MA, May 9.

———. 2016b. E-mail communication, September 5.

Charo, Alta and Henry Greely. 2015. "CRISPR Critters and CRISPR Cracks." *American Journal of Bioethics* 15: 11–17.

Collins, James P. 2016. Ullman Professor of Natural History and the Environment, Arizona State University. Telephone, August 8.

Davis, Joe. 2016. Resident bioartist, George Church Lab, Harvard University. On-site, Cambridge, MA, May 10.

Esvelt, Kevin M. 2016a. Associate Professor, MIT Media Lab, MIT. On-site, Boston and Cambridge, MA, May 10–11.

———. 2016b. E-mail communication, August 30.

———. 2016c. "Gene Editing Can Drive Science to Openness." *Nature* 534: 153.

Esvelt, Kevin et al. 2014. "Concerning RNA-guided Gene Drives for the Alteration of Wild Populations." *eLife* 3e03401: 1.

FDA. 2016. "U.S. Food and Drug Administration, Oxitec Mosquito." Available at: www.fda.gov/AnimalVeterinary/DevelopmentApprovalProcess/GeneticEngineering/GeneticallyEngineeredAnimals/ucm446529.htm

Friends of the Earth. 2016. "Genetic 'Extinction' Technology Rejected by International Group of Scientists, Conservationists and Environmental Advocates." *News Release*. September 1. Available at: www.foe.org/news/news-releases/2016-08-genetic-extinction-technology-rejected-by-international-group-of-scientists

Harmon, Amy. 2016. "Panel Endorses 'Gene Drive' Technology That Can Alter Entire Species." *N.Y. Times*. June 8. Available at: www.nytimes.com/2016/06/09/science/national-academies-sciences-gene-drive-technology.html

James, Anthony A. 2016. Distinguished Professor of Microbiology & Molecular Genetics, University of California, Irvine. Telephone, August 16.

Kohn, Eduardo. 2013. *How Forests Think: Toward an Anthropology Beyond the Human*. Oakland, CA: University of California Press.

Langston, Erica. 2016. "Voters in This Florida County Just Approved GM Mosquitoes to Fight Zika." *Mother Jones*. November 16. Available at: www.motherjones.com/environment/2016/11/key-haven-monroe-county-florida-gm-mosquitos-2016-election

Latour, Bruno. 1993. *We Have Never Been Modern*. Translated by Catherine Porter. Cambridge, MA: Harvard University Press.

Ledford, Heidi and Ewen Callaway. 2015. "'Gene Drive' Mosquitoes Engineered to Fight Malaria." *Nature News*. November 23. Available at: www.nature.com/news/gene-drive-mosquitoes-engineered-to-fight-malaria-1.18858

Lezaun, Javier and Natalie Porter. 2015. "Containment and Competition: Transgenic Animals in the One Health Agenda." *Social Science and Medicine* 129: 96–105.

Nading, Alex. 2014. *Mosquito Trails: Ecology, Health, and the Politics of Entanglement*. Oakland, CA: University of California Press.

National Academies of Sciences, Engineering, and Medicine. 2016. *Gene Drives on the Horizon: Advancing Science, Navigating Uncertainty, and Aligning Research with Public Values*. Washington, DC: The National Academies Press.

Novak, Ben. 2016. Lead Researcher, the Great Passenger Pigeon Comeback, Revive & Restore, The Long Now Foundation. Telephone, January 28.

Oxitec. 2016. *Oxitec's Vector Control Solution: A Paradigm Shift in Mosquito Control*. Available at: http://cdn.oxitec.com/wp-content/uploads/Oxitecs-Vector-Control-Solution-A-Paradigm-Shift-in-Mosquito-Control.pdf

Oye, Kenneth A. 2015. "Regulating Gene Drives." *Science* 345: 626–628.

Powell, Alvin. 2016. "Deploying Mosquitoes Against Zika." *Harvard Gazette*. March 28. Available at: http://news.harvard.edu/gazette/story/2016/03/deploying-mosquitoes-against-zika

Sankar, Pamela L. and Mildred K. Cho. 2015. "Engineering Values into Genetic Engineering: A Proposed Analytic Framework for Scientific Social Responsibility." *American Journal of Bioethics* 15 (12): 18–24.

Smidler, Andrea. 2016. PhD student, Harvard School of Public Health. On-site, Boston, MA, May 9.

Smidler, Andrea, John Min and Kevin M. Esvelt. 2016. "Harnessing Gene Drive Systems." In *Roadmap to Gene Drives* (draft, cited with permission).

Tomasik, Brian. 2015. "The Importance of Wild-Animal Suffering." *Relations* 3: 134–152.

Webber, Bruce, S. Raghu and Owain R. Edwards. 2016. "Opinion: Is CRISPR-based Gene Drive a Biocontrol Silver Bullet or Global Conservation Threat?" *Proceedings of the National Academy of Sciences* 112: 10565–10567.

Part II

Technologies of
Governance

Chapter 4

Laws of Containment

Control without Limits in the New Biology

J. Benjamin Hurlbut

Recent advances in genetic engineering techniques, particularly those associated with CRISPR, have been hailed as offering extraordinary new powers to manipulate cells, organisms, and ecosystems. *Science Magazine* has named CRISPR the 2015 breakthrough of the year, declaring a "CRISPR Revolution." The barriers that CRISPR broke through were not only technical, but also social and ethical. *Science* notes that its most spectacular applications, in particular the first deliberate editing of the DNA of human embryos, have "roiled the science policy world." CRISPR, according to the magazine, is revolutionizing not just the material foundations of life, but social, moral and political life. "For better or worse," *Science* declares, "we all now live in CRISPR's world" (Travis 2015).

The advent of CRISPR surely represents an important moment in the history of modern biology. Yet the language of radical novelty and revolution are entirely familiar. Advances in contemporary bioscience and biotechnology are often marked (and celebrated) as revolutionary: novel powers over life that portend profound changes to human lives and societies. At the same time, associated risks are rarely described in similar terms. Technological potentials are revolutionary, unlimited, and radically novel; risks are defined, knowable, and susceptible to well-established techniques of assessment and management. The asymmetrical treatment of potentials for benefit and risk is systemic and has characterized biotechnology since the development of recombinant DNA in the 1970s. It extends beyond the rhetoric of hype into authoritative scientific, regulatory, and political discourse around biotechnology, and is reflected in, and gives shape to, agendas of bioscience research and biotechnology governance in the United States.

Risks tend to be approached in circumscribed terms. They are defined according to an authoritative scientific assessment of what technological advances are plausible and what known risks might accompany them and are approached as problems of containment. In general, it is only once a technological scenario is deemed plausible and potential risks and "unintended consequences" are identified that ethical evaluation of "impacts" and "consequences" begins and costs and benefits are weighed. By contrast, biotechnology's potential for generating revolutions and "solutions to major societal needs" (National Research Council 2009) tend to be framed in open-ended terms, with the corollary imperative that science be set free to realize them. For instance, the U.S. Bioeconomy Blueprint, prepared by the Obama Administration, points to trends and technical advances in the biosciences, including "technologies not yet imagined" that "foreshadow major advances" in health, agriculture, energy, and manufacturing (U.S. White House 2014). Risk, safety, and ethical concerns are, by contrast, treated as knowable, manageable, and often a consequence not of technological innovation, but of "misuse" (U.S. White House 2014; cf. Doezema and Hurlbut forthcoming).

The revolution–risk asymmetry has been an important element in a reorientation within the biosciences from natural history to engineering and from analytic to synthetic biology (Carlson 2010). This is a reorientation not only in technique, but in the self-concept and ethos of a growing community within the biological sciences (Endy 2005). It is reflected, for instance, in the technological aspirations of what has been called the "New Biology," a phase of the biological sciences in which they "converge" with engineering, computing, and information sciences (Sharp et al. 2011). The New Biology, on this account, holds extraordinary technological potential, and thus will be capable of "enunciate[ing] and address[ing] broad and challenging societal problems" (National Research Council 2009, 3). At the core of this vision is the imperative to intervene in life: no longer merely focused on expropriating the products of natural evolution to the controlled, experimental environment of the laboratory, the idea of the New Biology is grounded in the presumption that its creations will necessarily exit the laboratory and inhabit bodies, industrial systems, and ecosystems. This presumption of release is linked to a corollary promise of containment: technologies to be released will nevertheless remain controlled and governed.

The revolution–risk asymmetry figures centrally in the New Biology, but is not new. It has shaped biotechnology governance since the mid-1970s. In this chapter, I offer a partial genealogy of this asymmetry and explore its implications for the contemporary governance of biotechnology. I focus in particular on the development of the notion of "biological containment." Since the inception of biotechnology, biological containment has been offered as an approach to governance that could be engineered into living systems, thereby privileging a narrow conception of risk. Biological containment is typically thought of as a biotechnological mechanism, but it is more appropriately described as a framework or regime of governance. From its inception, it represented a confidence and presumption that living entities could be engineered to contain themselves by intrinsically limiting their own capacities of survival. Thus, biological containment was offered—and continues to be offered—as an adequate means of governance, even in the absence of specific biotechnical tools that could achieve its promise in practice. The tools were not the foundation for governance; rather, the vision of governability—and its codification in a containment-focused regime of governance—was the warrant for constructing the tools.

Three developments trace back to the inception of biological containment as a framework for governance: first, a normalized approach to governing biotechnology with a stylized mode of constructing and evaluating "risks" in technical terms and as technology-specific problems of containment (an approach with a precedent in nuclear energy, but that is evident in many domains of U.S. science and technology; see Jasanoff and Kim 2009, 2013); second, the normalization and intensification of a tendency to describe bioscientific projects as having broad technological potential (and thus social benefit); and third, a politics of agenda setting that defers to scientific declarations of what is possible and plausible as a prerequisite to engaging in normative evaluation and governance. This chapter offers a genealogy and analysis of biological containment, demonstrating the central role that this technical concept has played in an imaginary of the unlimited potential and governability of biotechnology, and in the institutionalization of corollary practices of governance (Hurlbut 2015; Jasanoff 2015).

Containing Life

The development of molecular biology was predicated on the promise of containment. Anxieties about risk and safety at the advent of recombinant DNA in the 1970s were allayed by assurances that the environment of the laboratory and the techniques employed therein would guarantee that its products could not escape into the wider world. One of the key innovations of that period was the idea of biological containment—i.e., of engineering living systems to be dependent upon laboratory conditions, such that were they to escape that environment, they would not survive (Curtiss III et al. 1977).

The idea of biological containment both affirms and allays the worry that regulatory rules and laboratory environments have gaps. These human-managed systems to keep bioengineered life forms locked up can fail. Biological containment, by contrast, builds law and order into the living systems themselves, thereby engineering out reliance on brittle rules and fallible human practices. This is not only a technical strategy, but also a jurisdictional one. By rendering governance "intrinsic" to the living entity itself, its designers can claim to have eliminated any risk of losing control and thus any need for external oversight. At the same time, they claim a kind of unlimited jurisdiction over matters of risk: where a given biotechnology travels, so too will the apparatus of governance built into it, thereby trumping, displacing, and obviating the need for social regimes of control. In this sense, the idea of biological containment functions as an autochthonous repertoire of governance within the biosciences, deployed to do the work of (and forestall the construction of) conventional law. By treating biological systems as intrinsically controllable (but see Chapter 9, this volume), it underwrites the notion that they are governed most effectively if governance is approached as a matter of technical expertise.

Biological containment has long figured in the governance of biotechnology. The concept was one of the signature ideas that emerged out the 1975 Asilomar meeting on recombinant DNA. The meeting was convened by leading figures in molecular biology to lift a self-imposed moratorium on recombinant DNA experiments that were thought to be potentially hazardous (Wright 1994). The moratorium and call for self-imposed regulation elicited strong feelings from within the scientific community; some scientists were supportive of the moratorium, but many were highly critical. For critics, it threw into question the norm of scientific freedom that (putatively) governed the biological sciences. A figure no less prominent than James Watson, Nobel laureate and co-discoverer of the structure of DNA, declared at Asilomar that externally imposed rules not only should not, but also *could not* realistically reach into the laboratory, because researchers would do as they pleased behind closed doors. To his scientist colleagues he declared, "there should be no illusions that regulation is possible" (quoted in Weinberg 1975). The organizers, led by Stanford biochemist Paul Berg, saw things differently. It was, in their view, imperative that the Asilomar meeting produce a foundation for scientific self-regulation, both to ensure safety and to forestall the threat of formal, governmental regulation. Asilomar was, in this respect, not only an expression of scientific responsibility, but also of control. The group of scientists who gathered at Asilomar claimed the authority to define the parameters of possibility for this emerging scientific territory (Hurlbut 2015).

The meeting produced a set of recommendations for lifting the moratorium. The recommendations were grounded in the promise of containment: the promise that biohazard risks could be matched with laboratory practices that would in turn ensure that no dangerous product of the laboratory would escape into the wider world. Technical containment was also a means for containing the threat of external intervention by policy makers and public into laboratory science. In asserting that their program of containment was sufficient to address all risks associated with recombinant DNA (rDNA), the Asilomar scientists were also justifying the exclusion of the wider public from the decision-making process.

The rules for containment that were developed at Asilomar focused on constructions of risk that were narrowly technical and governable within the laboratory. In this respect, the focus on laboratory risk reflected and codified the parameters of conversation at the meeting. In opening the 1975 meeting, organizing committee member and soon-to-be Nobel laureate David Baltimore delineated the narrow focus of conversation by declaring two topics out of bounds: the biosecurity implications of rDNA and its social and ethical implications (Wright 1994). The former he marked as a problem for the government and the latter as a matter of private moral and religious beliefs. Thus, instead of assessing how rDNA might be used, and what those uses might mean for wider society, the meeting focused narrowly on technical risk assessment. That technical framing was used not only to justify the recommendations that came out of the meeting, but also to legitimize the lack of public participation in formulating them. If risk could be contained to the laboratory, why should the wider public be involved?

The exclusion of policy makers and the wider public from the Asilomar meeting would become a cause of significant controversy. In 1976, Senator Edward Kennedy (D-MA) criticized the Asilomar scientists for appropriating the authority to govern: "they were making public policy, and they were making it in private" (quoted in Culliton 1975, 1188). In the two years following Asilomar, there was a significant political struggle over whether rDNA research would be subject to more than scientific self-regulation. Recombinant DNA was the subject of heated political debate from Capitol Hill to town halls in cities like Cambridge, Massachusetts and Berkeley, California. At the heart of these debates was the question of whether scientific experts could legitimately address questions of policy that necessarily implicated the public and the environment. Figures like Kennedy demanded processes of participation and judgment that comported with democratic norms.

Scientists responded by asserting that the only relevant questions regarding governance of rDNA research were technical in nature, and that the best experts had assessed the risks and established adequate measures for containing them. They also began to assert more vocally and vigorously that rDNA held untold potential for generating social benefits—benefits that would be compromised if policy makers did not themselves take steps to contain public fears. For instance, in 1977, 137 scientists signed an open letter to Congress predicting that "the benefits of recombinant DNA research will be denied to society by unnecessarily restrictive legislation" (Gilbert 1977). A few months later, National Academy of Sciences president Philip Handler testified to Congress, quoting from a recent National Academies of Sciences, Engineering, and Medicine report:

> Recombinant DNA research under the NIH guidelines has such great promise for rapid future benefits and so little chance of causing harm that its regulation should be implemented with a minimum of restriction. Cumbersome and punitive legislation is not needed. The financial cost of overly cautious containment and enforcement, the delay in achieving benefits and the penalties incurred by restricting freedom of inquiry are real risks to be considered in setting up regulations.
>
> (U.S. House, Committee on Science and Technology 1978)

Yet the approach to physical containment that was developed at Asilomar was vulnerable to two objections: first, the possibility that the experts might underestimate the risks associated with particular experiments and thus fail to subject them to adequate containment measures; and second, that, as Watson suggested, the researchers themselves might break the rules. Sydney Brenner, an accomplished scientist who was also a member of the Asilomar organizing committee, suggested a solution to these vulnerabilities: containment could be guaranteed by engineering safety mechanisms into the experimental biological entities themselves. This idea of biological containment—to be distinguished from physical containment systems that would simply attempt to prevent potentially dangerous organisms from escaping—became a centerpiece of the Asilomar recommendations. It represented "the most significant contribution to limiting the spread of recombinant DNA" because "physical containment can reduce, but not eliminate, the possibility of spreading potentially hazardous agents" (Berg et al. 1975). Thus, the centerpiece of the recommendations were technologies of governance that could only be generated by bioscience itself.

Importantly, biological containment was put forward as a mode of governance rather than as a specific tool or technology. Although ideas were offered at Asilomar about how biological containment might be achieved, the technical means did not yet exist. Mechanisms were developed within a couple of years, the most important of which was Roy Curtiss's chi 1776, an enfeebled version of *E. coli* K12 that relied on specific laboratory conditions to be viable (Curtiss III et al. 1977); within a few years, other tools also emerged (for instance, Botstein et al. 1979). The concept of biological containment, however, and the role that it played in the governance of rDNA research, went well beyond chi 1776: it was advanced as a regulatory regime, a self-sufficient strategy for delineating, containing, and eliminating risk. Specific biological tools were slotted into a role that the concept of biological containment described in general terms. The idea of containment thus traded on the notion that risk could be treated as an engineering problem, and that the difficulties of legislating and enforcing mechanisms of risk management could be overcome by incorporating law and order into biotechnology itself. In figuring containment as the sole problem of governance, and biological containment as the most reliable mechanism to address it, governing science became a task of engineering organisms. This task relied entirely on expert vetting and certification of technologies of biological containment. The consequence was that scientific evaluation of the reliability of built-in mechanisms of control became the test for whether there was any need to ask additional questions about risks and benefits or expand the scope of deliberation. Insofar as the living systems could be made to limit themselves, this also seemed to eliminate legitimate reasons for public oversight.

The central point I wish to make here is that by imagining risks as problems of containment, and containment as primarily a matter of engineering safe biological systems, the range of questions that could be asked about the potential for harm was significantly truncated. My intent is not to question the efficacy of particular technologies of biological containment like chi 1776, but rather to observe that the question of whether those technologies are effective have, to a large degree, come to stand in for questions of how to govern scientific research. The focus on containment, and on biological containment in particular, has played an important role in confining evaluations of risk to narrow, technical questions focused on specific experiments rather than on long-term technological trajectories or more comprehensive evaluation of regulatory aims and approaches. The effect has been to build a significant asymmetry in constructions of potential risks and benefits into practices of governance, treating risks as knowable, manageable, and solvable short-term technical problems that are circumscribed to the laboratory, and benefits as revolutionary, open-ended, and unlimited in potential benefit to society.

In short, the promise of biological containment as articulated at Asilomar was that only benefit, and not risk, would migrate from laboratory to world. As such, the idea has helped to underwrite the asymmetrical treatment of risks and benefits in governance of biotechnology: risks are treated as technical problems amenable to technical solutions and thus as internal to, and solvable by, science; by contrast, wider society is invited to participate in enthusiastic imaginations of beneficial technological futures. Thus, insofar as the aim of this chapter is to invite skepticism about the adequacy of dominant modes of governing biotechnology, the object of critique is not the epistemic claims about whether particular tools of biological containment do what their designers claim they do; it is, instead, the work that the idea of biological containment itself does in shaping the contours of deliberation and governance.

Control without Limits

The risk–revolution asymmetry has only intensified over the course of the four decades since Asilomar. During this period, biotechnology innovation has proceeded by presuming that its products will eventually be released from the laboratory into the wider world. At the same time, the promise of containment persists, even as mechanisms for achieving it have become progressively more complex and speculative. In the 1970s, biological containment was conceived as a mechanism for ensuring that inadvertent release of engineered organisms into the environment would not pose a risk. Four decades later, it is deployed to justify intentional release. Biological containment offers, in effect, a strategy to keep engineered organisms "isolated from natural ecosystems" (Mandell et al. 2015), even as they are released into the living world. Synthetic biologists have imagined a menagerie of biological mechanisms for achieving such isolation-with-integration, including techniques like auxotrophy, gene-flow barriers, "kill switches," artificial genetic codes, and metabolic dependence on artificial amino acids.

These tools of containment are "experiments of reassurance" (Rabinow and Bennett 2012), the counterpart to synthetic biology's commitment to "democratizing" bioengineering. Proponents of synthetic biology take for granted that the tools of bioengineering should be made widely available with low barriers of entry for "playing around" with living systems (Endy 2005; Kuiken, Chapter 5, this volume). The concept of biological containment offers a strategy for engineering in responsibility by delegating the task of governance to biological parts built into the organism itself that will ostensibly function as technologies of total containment.

At the same time, biological containment is seen as largely removing risk from the picture of a world transformed by biotechnology, and thus as authorizing an agenda for innovation that presumes that new life forms will necessarily move from the laboratory into the environment. As one review of biological containment strategies in synthetic biology puts it: "if synthetic organisms and their derivatives are to become as ubiquitous as electronic devices, then synthetic biologists must openly address the responsible and safe use of synthetic biological systems" (Moe-Behrens, Davis, and Haynes 2013, 2). This straightforward transition from "if" to "then" is built on the scaffolding of a contingent logic. The presumption that the world should and will become populated with novel life forms requires (i.e., authorizes) synthetic biologists to define, address, and guarantee the safety of these systems.

There is a politics of jurisdiction in the idea of "preparing synthetic biology for the world" (ibid.). Techniques of biological containment are intended not only to contain biohazard risks, but also to contain the threat of public reaction. Negative assessments that:

> cast synthetic biology as a threat to human well-being . . . diminish the fact that the core ethos of synthetic biology . . . is a design process that aims to make human inventions reliable, predictable and safe . . . synthetic organisms might turn out to be the best solution for global health challenges and ecological problems.
>
> (ibid., 2)

Yet such assessments indicate "how the public may react when coordinated efforts towards executing containment and control strategies are not highly visible" (ibid., 2). Biological containment is offered as a means to facilitate synthetic biology's project of generating new biological entities and introducing them into the world, to quiet public calls for governance by appearing to eliminate risk, and to displace ideas of "tampering with nature or 'playing God'" by maintaining an apparent separation between "contained" engineered organisms and natural environments (ibid.; cf. Presidential Commission 2010). As such, synthetic biologists claim the competency and thus the authority to govern their own creations. In the same move, they assert the authority to define the terms of public debate about risks and benefits by reducing risks to solvable engineering problems. And, by constructing risk as a solvable by technical but not by social means, synthetic biology construes legal and political institutions as highly limited in their capacity to manage risk without inhibiting beneficial technological innovation. By contrast, "intrinsic" containment appears to have a kind of unlimited jurisdiction by virtue of travelling with the technology itself. Efficacy, monitoring, and enforcement are all enacted within and by the organism itself as it executes its self-limiting genetic instructions.

Recent advances in gene editing, and CRISPR in particular, dramatically expand opportunities for reengineering life. Recognition of this potential has rightly elicited a reappraisal of the modes of governance that have emerged in conjunction with biotechnology. While the notion that technological innovation outpaces regulation is an oft-invoked truism (and is not born out by actual practices of regulation), with gene editing there is an unusual uneasiness about approaching governance in the same old ways. For once, there appears to be a certain openness to seeing the risks as well as the benefits as revolutionary (see Esvelt in Chapter 1, this volume, for examples of such openness, although they are in this case joined with a doubling-down on scientific confidence in biological containment as an essential strategy of governance).

Yet debate continues to cleave to containment as a mode of governance. The same repertoire of techniques that might produce unprecedented risks are also available for innovating mechanisms of biological containment. This paradox is particularly evident in the case of gene drives, which are designed to spread specific genetic changes throughout a natural population until it is transformed or destroyed. Because gene drives expand infinitely by design, any unforeseen gaps in containment would make a local test into a global intervention. Biological containment mechanisms have been offered as a means to tame this powerful technology, rendering global gene drives "local" by building in mechanisms of self-attenuation, or by engineering "reversal drives" to undo the "unintended consequences" of intentionally released gene drives. There is some disagreement about what it would take to keep a gene drive contained, but, by and large, that disagreement is technical and treats containment as a design problem (Esvelt, this volume; National Academies of Sciences, Engineering, and Medicine 2016). Once again, this disagreement reflects the canonical moves associated with biological containment: governance can be engineered-in, such that the technology will mitigate its own riskiness. And once again, the narrow construction of risk displaces questions that transcend immediate applications: questions about what the technology is for— what future it contributes to, what good it serves, and what "risks" the technological project might pose to established natural, social, and moral orders.

The U.S. National Academies of Sciences, Engineering, and Medicine's 2016 evaluation of gene drives has recommended field trials under certain circumstances, namely circumstances in which there is adequate scientific confidence in mechanisms of containment. However, the report also notes the need to attend to diverse, value-laden, public views about risks and benefits before undertaking any intentional environmental release. But there is a tension in the report's treatment of the relationship between expert judgment about "how much" risk is posed by release and its affirmation of public deliberation as a necessary ingredient. This tension reflects a prior, well-entrenched distinction between containment-based governance as a technical design problem and the political question of how societies should make judgments about the (un)desirability of technological projects. Affirmations of "public engagement" notwithstanding, an imaginary of containment delimits the terms of debate: public deliberation is required only where there is uncertainty about technical capacities for containment. Thus, "public engagement" remains subject to prior jurisdictional commitments that limit the public's role. Where "uncertainty" is offered as the warrant for seeking public input about what is at stake, modes of biological containment that purport to reduce or eliminate such uncertainties are offered as both necessary and sufficient to quiet wider public concerns.

This jurisdictional relationship is also an epistemological one: confidence in mechanisms of biological containment underwrite experts claims to "know" that a public concern is unrealistic and therefore unwarranted—not because underlying values are misguided, but because mechanisms of containment ensure that they simply will not be contravened (cf. Braverman, Chapter 3, this volume). Conversely, to be heard, value-laden concerns get channeled into expressions of skepticism about whether mechanisms of containment are truly fail-safe. These relationships are not tied to particular epistemic or moral disagreements. They are relational and jurisdictional; they precede and configure those disagreements.

Political Risks

The corollary to the promise that risk can be contained is the promise of an end-less frontier of technological progress. It has become normal, obligatory even, to preface any discussion of an emerging biotechnology, whether laudatory or critical, with a statement about its "tremendous potential" for benefiting society. In public discourse, in the United States at least, benefits are assumed, but risks have to be demonstrated. Risk assessment has become a technical enterprise, and where contested, a battleground for expert credibility. More overtly, moral idioms tend to be excluded, a tendency that is underwritten by the notion that values are parochial whereas scientific knowledge and its technological incarnations are universal (Jasanoff 1999). In this imaginary of the knowledge–norms relation-ship, the reach of values is limited to the legal and political authority of the actors who espouse them, be they individuals, institutions, or nation states. On the other hand, the jurisdictional reach of knowledge and technology is unlimited except where it encounters prohibition or denialism.

I have argued that biological containment as a mode of governance is one pat-tern of political practice through which imaginations of risk in biotechnology have been narrowed in scope such that they are susceptible to technical evaluation and remediation. As the aperture of risk analysis has narrowed, the presumed reach of its repertoire has lengthened. Yet this arrangement is not inevitable or given in advance; it is grounded in a jurisdictional relationship that was itself a political achievement in the early days of biotechnology. To illustrate this, I briefly return to this history.

By 1979, political controversy surrounding rDNA had largely subsided. Policy makers and publics had generally accepted scientists' assurances that the tech-niques were safe. The Asilomar recommendations had been formalized as guide-lines that governed publicly funded rDNA research, and the Recombinant DNA Advisory Committee (RAC) had been tasked with enforcing them. In 1976, the preamble to the guidelines was amended to read:

> The experiments now permitted under the guidelines involve no known addi-tional hazard to the workers or the environment beyond the relatively low risk known to be associated with the source materials. The additional hazards are speculative and therefore not quantifiable. In a real sense, they are consider-ably less certain than are the benefits now clearly derivable from the projected research.
>
> (U.S. House, Committee on Science and Technology 1978)

Apologists for rDNA research increasingly linked their assurances of safety to warnings that if the public were to intervene and set limits, it would be denying itself untold future benefits.

Yet the apparatus that governed research was tenuous. A science writer embed-ded in the laboratory of Herbert Boyer in the late 1970s observed numerous violations of the rules, both inadvertent and intentional (Powledge 1977). Boyer, who was awarded the patents for fundamental rDNA techniques, would go on to team up with Stanley Cohen to found Genentech, the world's first and most successful biotechnology company. As these academic scientists made this unusual migration into industry, they crossed over into an ungoverned territory. Although Congress had attempted to apply public science's rules to private sector research, the overwhelming opposition of the scientific community to any formal regulation killed this effort. Even as National Institutes of Health Director Donald Fredrickson was leading the effort to forestall legislation, he had convinced Roy Curtiss to seek a patent for chi 1776. His idea was to use private intellectual property as an instrument of public policy, making industry adherence to the guidelines a condition of licensing.

By the time the U.S. Supreme Court declared Ananda Chakrabarty's oil-eating microbe an invention, thus clearing the way for Curtiss to receive a product patent for chi 1776, the RAC had made the bacterium moot (U.S. Patent No. 4,190,495 1980). In 1979, after a reshuffling of RAC membership, a proposal was floated to downgrade experiments involving *E. coli* K12 to the lowest level of containment. Doing so would effectively reclassify K12 as providing adequate biological containment, thereby largely obviating the need for chi 1776. The rationale for the proposed change was based on a small body of research that suggested that *E. coli* K12 could not survive in the human gut or in the environment, therefore in itself exhibiting a *de facto* system of biological containment. The RAC's decision turned not only on the question of whether K12 could survive outside the laboratory, but whether bacteria engineered to express an antigen or protein that did colonize a person's gut would put them at risk. In facing this question, the body had to contend with a dilemma about what one member referred to as the "Brenner principle": the notion, attributed to Sydney Brenner, that placing restraints on research is legitimate only when a hazard is clear, imminent, and scientifically known. The effect of this principle was to exclude any "imaginary hazard" (Henig 1979). By limiting the parameters of legitimate risk assessment to what is known, known unknowns (and, of course, unknown unknowns) were marked as speculative and therefore inappropriate for science-based evaluation. Echoes of the Brenner principle persist in the coordinated framework's commitment to "sound-science" based regulation. Yet this is no less a political commitment today than it was in 1979, for it amounts to a commitment to narrow the scope of imagined futures to the possibilities that authorized experts deem plausible and quantifiable and to mark all else as speculative. Indeed, the corollary to the construction of risk that containment privileges is the normative commitment to privileging imaginations of progress over a politics of precaution. "Policies based on the precautionary principle could stunt the development of synthetic biology" (Moe-Behrens, Davis, and Haynes 2013).

One of the key moments that turned the political tides of the rDNA controversy was a meeting at Falmouth, MA in June 1977. At that meeting, a number of the scientists who were present at Asilomar came to the collective conclusion that earlier worries about the risk of producing pathogenic *E. coli* had been overblown (Gilbert 1977). A report later prepared by the RAC referenced this consensus in asserting that "strain K-12 cannot be made pathogenic even when provided . . . with the genes for known toxins and other pathogenic properties." A researcher who participated in the Falmouth meeting recently told me that the Falmouth group came to this conclusion after they could not think of a way to render *E. coli* pathogenic by design and therefore judged that the risk of doing so inadvertently was minimal. "Ten years later," he added, "I could have [made K12 pathogenic] with both hands tied behind my back."

These moments illustrate an unsurprising (even if inconvenient) truth: capacities of governance are indexed to the modes of imagination that inform them and which they, in turn, discipline. The wildly optimistic visions of the biotechnological future that have come to be so ubiquitous that invoking them is virtually obligatory are joined to highly circumscribed imaginations of risk. Those imaginations, in turn, depend upon the appearance of universality and unlimited jurisdiction associated with authoritative scientific knowledge, and upon a corollary promise of technological precision and control. Yet the completeness of the picture turns out upon inspection to be patchy and riddled with gaps and built upon moves that occlude them. Some moves consist in parochial commitments to an imaginary of knowledge as governance; others exploit the limited jurisdiction of political authority to avoid being subjected to the test of sustained public scrutiny. Both kinds of moves have been essential to biological containment as a framework of governance.

The AquAdvantage salmon offers an illustrative example of such moves. After nearly two decades of review, in 2015 it was the first genetically engineered animal to be approved by the United States Food and Drug Administration (FDA) for human consumption. This genetically engineered Atlantic salmon can grow to market size about twice as fast as its conventionally bred cousins. Even though it was produced over two decades ago, the fish has widely been seen as a test case for regulation of genetically engineered animals—an agricultural category that is ballooning with the advent of CRISPR because this technology makes the genetic modification of animals far easier.

AquAdvantage has been celebrated in all the standard terms. It has been posed as a "revolution" in aquaculture, a solution to overfishing and depleted stocks, and an early example of the kinds of food productivity that biotechnology will supply for a growing global population. At the same time, it offers abundance without risk. In principle, were it to escape into the wild, the salmon could represent a significant environmental risk to its wild relatives. But good governance has been engineered into the salmon in the form of biological containment (the fish are all female and sterile), thereby untethering environmental security from nationally specific rules with potentially incomplete enforcement and ensuring that the fish can safely circulate to global markets without risk to natural populations.

Yet the promise of unlimited jurisdiction is belied by the baroque regulatory arrangements on the basis of which FDA granted approval. In evaluating the

AquAdvantage salmon, the FDA simplified its regulatory responsibilities by displacing environmental risks to other jurisdictions. AquaBounty is based in the United States, the eggs are produced and fertilized in Canada, and the fish are grown to size in Panama. This transnational production regime absolves the FDA of the obligation to assess the potential that containment measures could fail and thus impact the environment. Because "the areas . . . most likely to be affected . . . lie largely within the sovereign authority of other countries," environmental effects were not evaluated "except insofar as it was necessary to do so in order to determine whether there would be significant effects on the environment of the United States" (Center for Veterinary Medicine 2015). Production outside of the United States (and, thus, outside the jurisdiction of an FDA environmental assessment) was a requirement of approval. Notwithstanding the promise of governability through mechanisms of biological containment, risk was addressed in this context by locating it in other legal jurisdictions.

The advent of CRISPR has generated significant "regulatory uncertainty" about what rules will apply, for what applications, and with what jurisdictional reach. At the same time, long established habits are kicking in. For all its radical novelty, the risks of gene editing are being characterized as no different than (and thus no more risky than) old techniques. It is presented as a more efficient means of intra-specific, conventional plant and animal breeding, a means of livestock crossbreeding without the time-consuming process of multiple crosses (see Travis, Chapter 8, this volume). Even in territories as fraught as human germline modification, some have suggested that gene editing improves on natural sex because it poses less risk of mutagenesis (Harris 2015). Imaginations of a world transformed by gene editing run wild even as risks are domesticated to narrow, technical frames.

In December 2015, the United States National Academies hosted an International Summit on human gene editing. David Baltimore opened the meeting. Forty years earlier, he had opened the Asilomar meeting by limiting discussion to technical evaluation of rDNA risks. In his opening remarks at the 2015 Summit, he drew comparisons to the 1975 meeting, explaining why it was urgent to address the implications of gene editing, just as it had been urgent to the same extent with rDNA four decades earlier. Although genetic engineering techniques have been around for decades, he explained, they were imperfect, and applying them to humans was therefore "initially unthinkable." But with technological advances "the unthinkable has become conceivable." Thus, "now we must face the questions that arise: how, if at all, do we as a society want to use this capability?" (Baltimore 2015).

I have shown how in limiting the scope of discussion—and the imaginations of risk that were admissible in it—Asilomar pioneered an approach to the governance of biotechnology that systematically excluded public participation, and thus also eliminated the diverse wealth of moral imagination that it would offer. At the same time, the Asilomar scientists claimed to speak for society because they spoke for the future technological benefits that were promised to accrue to society. The idiom adopted at Asilomar has since been solidified and codified. There has been an attenuation in science governance of modes and structures for engaging imaginations of benefit and harm that are wider than those that tend to regulate biology (see also Newman, Chapter 7, this volume). Thus, Baltimore's call for a "we" who "must face the questions that arise" occludes the extent to which a robust "we" has been systematically displaced from the governance of biotechnology— displaced by the promise of robust forms of containment that ostensibly obviated the need (and thus, the obligation) to engage in robust processes of public deliberation and judgment.

Yet in the same sentence, Baltimore invoked the very dynamic that displaces that "we," the wider political community, from playing a central role in evaluating the "questions that arise" with advances in biotechnology. "The unthinkable has become conceivable." Excluding what scientific experts deem to be unthinkable from the deliberative democratic agenda, and inviting deliberation only once those experts deem it to be conceivable, has been a standard move since the inception of biotechnology. Indeed, the questions that "now we must face" have long been raised and long been set aside. For instance, in the early 1980s, the President's Commission—a public bioethics body in the United States—was called upon to explore the ethical implications of human germline genetic engineering. The Commission significantly narrowed the scope of its deliberations by refusing to reflect on scenarios that it deemed to be technically implausible or remote. It likewise rejected ethical discourse that emphasized radical disjunctions with the biological status quo, thinning discourse and narrowing discussion to applications that were deemed to be plausible, specific, and immediate (President's Commission 1982; cf. Evans 2002). This pattern has become normal and normative, particularly in bioethics discourse. Deference to scientific judgments about what is plausible and realistic and what is mere "science fiction" is a hallmark of mainstream bioethical evaluation of new technologies in the United States.

In this sense, the field of bioethics, which is supposed to provide an institutional reservoir of ethical capacity to address a deficiency of such capacity in the sciences, nevertheless buys into this foundational jurisdictional move. Scientific experts are authorized to declare what *is* (or is not) realistic, thereby delimiting the range of ethical questions that can be asked about what *ought* to be done (Hurlbut 2017). The recent statement on human gene editing from the Hinxton group—a small, ad hoc group of scientists, lawyers, and ethicists—offers an instructive example of this move. In the statement, democratic deliberation is treated as subsidiary to scientifically authorized judgments. The statement asserts that "a distinction should be made between objections that are based on technical or safety concerns and objections that reflect additional moral considerations." The statement continues:

> any constraint of scientific inquiry should be derived from reasonable concerns about *demonstrable* risks of harm . . . Policymakers should refrain from constraining scientific inquiry unless there is substantial justification for doing so that reaches beyond *disagreements based solely on divergent moral convictions.*
> (The Hinxton Group 2015; emphasis added)

This formulation treats technical judgments of safety as having greater force than democratic ambivalence about appropriateness. Disagreement and moral uncertainty are characterized as an inadequate warrant for democratically defined limits: until democratic judgment achieves universality and univocality, its jurisdiction is—and must be—limited.

Conclusion

I have argued that the idea of containment as a mode of governance has contributed to the narrow, technical approach to assessing potential risks associated with advances in biotechnology, with significant consequences for dominant modes of deliberation and governance of the biotechnological future. Governance-by-containment limits constructions of risk to unintended environmental or public health effects, treats risk management as primarily a technical matter of ensuring that "nothing bad gets out," and presumes that if all plausible risks have been adequately contained, environmental release is beneficial by definition. With biological containment, modes of governance that rely upon political judgment and social coercion are displaced by apparently more reliable biological mechanisms. Rather than rely upon fickle and fallible social instruments, the biotechnological product is made to ostensibly govern itself. Engineered organisms are thus "isolated" from natural environments, even as they are integrated into them, and concerns about the risks associated with radically altering the living world lose traction. Approached in this way, biological containment becomes the warrant for setting the New Biology free and for integrating biotechnology into the sociotechnical order: if risks have been excised from the nature of the entity itself, there is no justification for limiting the (presumptively) beneficial development of a technology, or its deployment into the wider world. Indeed, any externally imposed limits would themselves risk "delay in achieving benefits and the penalties incurred by restricting freedom of inquiry" (U.S. House, Committee on Science and Technology 1978).

Such modes of governance-by-containment reflect anxieties about progress, and about the potential that public ambivalence might stand in its way. I have sought to surface some of the moves and the habits of mind through which our institutions of governance (of which science is unequivocally one) displace ambivalence with a sense of clarity and confidence. Yet it is a clarity that is achieved by cleansing sociotechnical transformations of the ambiguities that inevitably accompany them. And it is a confidence that is grounded in a claim to a higher authority, one that purports to transcend the parochialism of ethical judgment with the certitude of knowledge.

Yet we would do well to refuse the promise of a smooth path and frictionless passage into the future. Confidence in the promise of containment amounts to little more than blindly believing that no harm will come. It is a reactive posture, one that responds only to the threat of harm, rather than being animated by an idea of the good. As such, it refuses the difficult task of confronting the question of the good, and engendering the capacities of reflection and deliberation necessary to address it. The task of achieving a "we" that is grounded in an orientation to the good is inhibited by a reactive posture. A posture that, by contrast, aspires to such a "we" celebrates the openness of the future and acknowledges the task of imagining and enacting it as a collective project. If the technologies of the moment truly offer the power to remake life, then they might also provide occasion to revisit and rewrite our programs of governance, and so too the habits of mind and modes of imagination that underwrite them. The present moment thus presents us with an opportunity—and with a responsibility—for innovation. This is an opportunity that should be acknowledged and seized upon.

References

Baltimore, David. 2015. "Human Gene Editing Summit: Context for the Summit." Washington, D.C., December 1. Available at: https://vimeo.com/album/3703972/video/149179797

Berg, Paul et al. 1975. "Summary Statement of the Asilomar Conference on Recombinant DNA Molecules." *Proceedings of the National Academy of Sciences* 72: 1981–1984.

Botstein, David et al. 1979. "Sterile Host Yeasts (SHY): A Eukaryotic System of Biological Containment for Recombinant DNA Experiments." *Gene* 8: 17–24.

Carlson, Robert H. 2010. *Biology Is Technology: The Promise, Peril, and New Business of Engineering Life.* Cambridge, MA: Harvard University Press.

Center for Veterinary Medicine. 2015. United States Food and Drug Administration Department of Health and Human Services. "Finding of No Significant Impact: AquAdvantage Salmon." Available at: www.fda.gov/downloads/AnimalVeterinary/DevelopmentApprovalProcess/GeneticEngineering/GeneticallyEngineeredAnimals/UCM466219.pdf

Culliton, Barbara J. 1975. "Kennedy: Pushing for More Public Input in Research." *Science* 188: 1187–1189.

Curtiss III, Roy. 1977. "Letter to Donald Fredrickson, Director, NIH." April 12. Available at: http://profiles.nlm.nih.gov/DJ/Views/Exhibit/documents/regulation.html

Curtiss III, Roy et al. 1977. "Biological Containment: The Subordination of Escherichia Coli K-12." In *Recombinant Molecules: Impact on Science and Society.* Edited by R.F. Beers and E.G. Bassett, 45–56. New York: Raven Press.

Doezema, Tess and J. Benjamin Hurlbut. Forthcoming. "Technologies of Governance: Science, State and Citizen in Visions of the Bioeconomy." In *Bioeconomies: Life, Technology and Capital in the XXIst Century.* Edited by Vincenzo Pavone and Joanna Goven. New York: Palgrave.

Endy, Drew. 2005. "Foundations for Engineering Biology." *Nature* 438: 449–453

Evans, John H. 2002. *Playing God?: Human Genetic Engineering and the Rationalization of Public Bioethical Debate.* Chicago, IL: University of Chicago Press.

Gilbert, Walter. 1977. "Recombinant DNA Research: Government Regulation." *Science* 197: 208.

Harris, John. 2015. "Why Human Gene Editing Must Not Be Stopped." *Guardian*, December 2. Available at: https://www.theguardian.com/science/2015/dec/02/why-human-gene-editing-must-not-be-stopped

Henig, Robin Marantz. 1979. "Trouble on the RAC: Committee Splits over Downgrading of E. Coli Containment." *BioScience* 29: 759–762.

The Hinxton Group. 2015. "Statement on Genome Editing Technologies and Human Germline Genetic Modification." Available at: www.hinxtongroup.org/hinxton2015_statement.pdf

Hurlbut, J. Benjamin. 2015. "Remembering the Future: Science, Law, and the Legacy of Asilomar." In *Dreamscapes of Modernity: Sociotechical Imaginaries and the Fabrication of Power*. Edited by Sheila Jasanoff and Sang-Hyun Kim, 126–151. Chicago, IL: University of Chicago Press.

———. 2017. *Experiments in Democracy: Human Embryo Research and the Politics of Bioethics*. New York: Columbia University Press.

Jasanoff, Sheila. 1999. "The Songlines of Risk." *Environmental Values* 8: 135–152.

———. 2015. "Future Imperfect: Science, Technology, and the Imaginations of Modernity." In *Dreamscapes of Modernity: Sociotechical Imaginaries and the Fabrication of Power*. Edited by Sheila Jasanoff and Sang-Hyun Kim. Chicago, IL: University of Chicago Press.

Jasanoff, Sheila and Sang-Hyun Kim. 2009. "Containing the Atom: Sociotechnical Imaginaries and Nuclear Power in the United States and South Korea." *Minerva* 47: 119–146.

———. 2013 "Sociotechnical Imaginaries and National Energy Policies." *Science as Culture* 22 (2) (June 1): 189–196.

Mandell, Daniel J. et al. 2015. "Biocontainment of Genetically Modified Organisms by Synthetic Protein Design." *Nature* 518: 55–60.

Moe-Behrens, Gerd H.G., Rene Davis, and Karmella A. Haynes. 2013. "Preparing Synthetic Biology for the World." *Frontiers in Microbiology* 4: 5.

National Academies of Sciences, Engineering, and Medicine. 2016. *Gene Drives on the Horizon: Advancing Science, Navigating Uncertainty, and Aligning Research with Public Values*. Washington, D.C.: The National Academies Press.

National Research Council. 2009. *A New Biology for the 21st Century*. Committee on a New Biology for the 21st Century: Ensuring the United States Leads the Coming Biology Revolution. Washington, D.C.: The National Academies Press.

Powledge, Tabitha M. 1977. "Recombinant DNA: Backing off on Legislation." *The Hastings Center Report* 7: 8–10.

Presidential Commission on Bioethical Issues in Biomedical Research. 2010. "New Directions: The Ethics of Synthetic Biology and Emerging Technologies." Washington, D.C.

President's Commission for the Study of Ethical Problems in Medicine and Biomedical and Behavioral Research. 1982. *Splicing Life: A Report on the Social and Ethical Issues of Genetic Engineering with Human Beings*. Washington, D.C: The Commission.

Rabinow, Paul and Gaymon Bennett. 2012. *Designing Human Practices: An Experiment with Synthetic Biology*. Chicago, IL: University of Chicago Press.

Sharp, Phillip et al. 2011. *The Third Revolution: The Convergence of the Life Sciences, Physical Sciences, and Engineering*. Cambridge, MA: Massachusetts Institute of Technology.

Travis, John. 2015. "Making the Cut." *Science* 350: 1456–1457.

U.S. House, Committee on Science and Technology. 1978. *Hearing: Recombinant DNA Act*. April 11. Washington, D.C.

U.S. Patent No. 4,190,495. 1980. "Modified Microorganisms and Method of Preparing and Using Same." Filed September 27, 1976. Issued February 26, 1980.

U.S. White House. 2014. "National Bioeconomy Blueprint." April 26. Available at: www.whitehouse.gov/administration/eop/ostp/library/bioeconomy

Weinberg, Janet H. 1975. "Decision at Asilomar." *Science News* 107 (12): 194–196.

Wright, Susan. 1994. *Molecular Politics: Developing American and British Regulatory Policy for Genetic Engineering, 1972–1982*. Chicago, IL: University of Chicago Press.

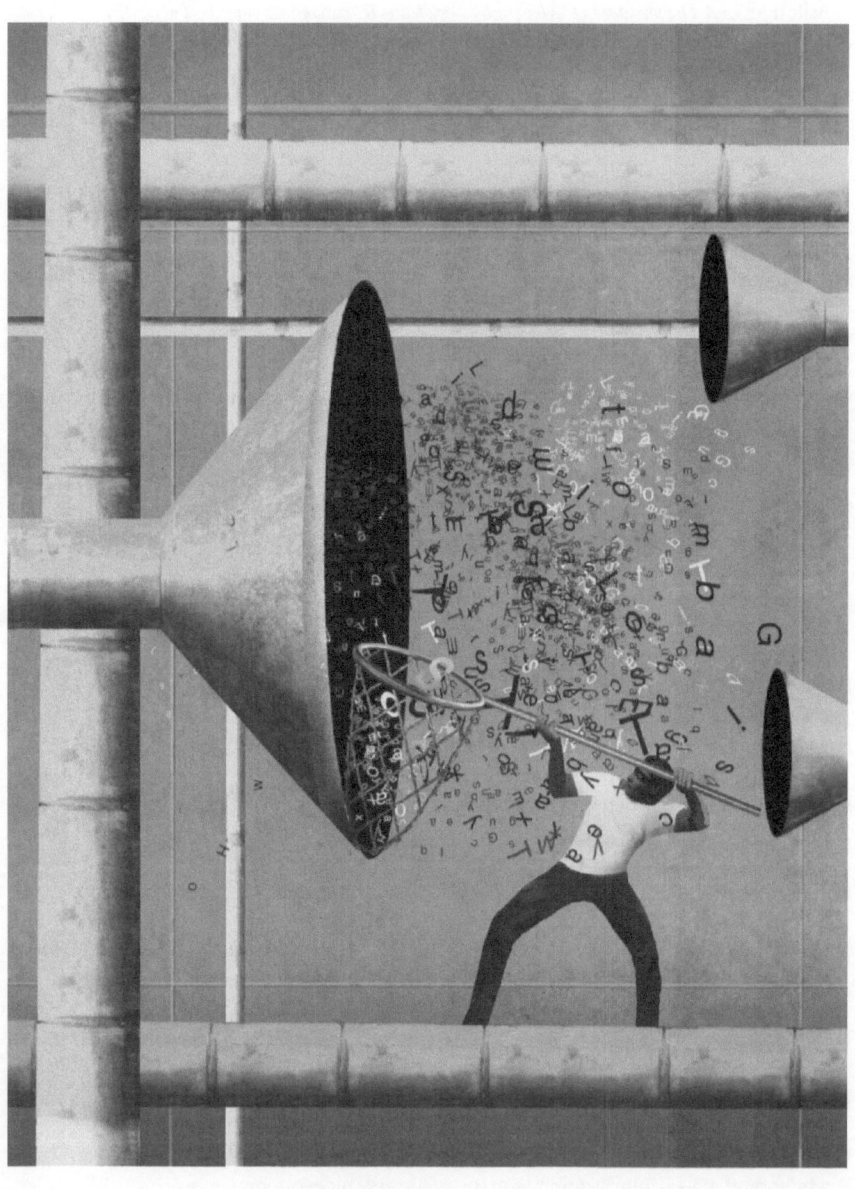

Chapter 5

Vigilante Environmentalism

Are Gene Drives Changing How We Value and Govern Ecosystems?

Todd Kuiken

Introduction

Vigilantes are members of self-appointed groups of citizens who undertake law enforcement in their community without legal authority, typically because the legal agencies are thought to be inadequate. Vigilantes resurrect images of the Wild West, where citizens take justice into their own hands in towns where no form of centralized governance has been established or enforced. In the past, some environmentalists used such vigilante tactics, which resulted in their labeling as "eco-radicals." Some used these tactics in order to save what they believe are sacred and important ecosystems. Are we beginning to see similar "eco-radicals" emerge from the conservation community with regard to gene drives? While it has been suggested that recent developments with CRISPR-Cas9 and gene drive technologies could address disease vectors, invasive species, and other conservation problems, these technologies are no longer restricted to traditional actors and are being developed within a governance and risk assessment void.

I feel conflicted about the use of gene drives for conservation. This confliction is rooted in my training as an environmental scientist and the environmental worldviews I hold, which often also stand in conflict with one another (i.e., deep ecology, eco-centrism, techno-centrism). In full disclosure, I have been closely working with groups that have been promoting and exploring the use of genetic technologies for conservation purposes. Additionally, in the past I have worked for large conservation organizations that tend to be averse to technologies like gene drives. While the views here are my own, I believe they reflect a complicated and evolving relationship between technology and its use in conservation and environmental protection.

This chapter will explore a series of questions related to the proposed uses of gene drives as a tool for conservation, including the philosophies behind these proposals, and issues of public trust and governance. It will not prescribe whether or not these are the "right" solutions for conservation. Instead, I would like to consider whether this biotechnological turn signifies a broader shift in the environmental community or if it is more aligned with a small niche group of eco-radicals.

Who Speaks for the Trees?

Much of the conflict around environmental protection stems from philosophy and religion. In contrast to what many religions teach, which is more focused on "transcending this world or obtaining divine rescue from it," Taylor suggests that those who adopt a spiritual view of nature believe that there is a "spiritual intelligence with whom one can be in a relationship" (Taylor 2010, 4). Taylor further argues that spirituality is about personal growth and understanding of one's place, which is "intertwined with environmentalist concern and action" (ibid., 3). Along these lines, an eco-centrist views humankind as part of a global ecosystem that is subject to ecological laws. These views and the demands of an ecologically based morality impose constraints on human action, particularly through limits on economic and population growth. Such proponents also hold a strong sense of respect for nature in its own right, as well as for pragmatic reasons, and typically advocate that if technologies are used, they should be democratized with little economic or political power (Pepper 1993). On the other hand, a techno-centrist worldview recognizes environmental problems, but believes that our current form of society will always solve them (through careful economic and environmental management), and still achieve unlimited growth. Such a techno-centrist worldview typically has little desire for genuine public participation in decision making (ibid.). Scarce articulates the tension between eco- and techno-centrism when he writes: "We have achieved what our forebears spent untold generations attempting to accomplish: Nature is ours. Now radical environmentalists want to give it back" (Scarce 2006, preface).

These and other environmental worldviews vary in degrees of care or responsibility placed on the individual and on society in relation to nature. For instance, "green religion" describes an individual's environmental behavior based on their perceived obligation, whereas a dark green religion perceives nature as sacred, as having intrinsic value, and therefore is due reverent care (Taylor 2010). Dark green religion emerges from deep ecology, first defined by Arne Naess in 1973. This worldview sees humans as just one species among many, having no special right to dominate or destroy the environment (Cramer 1998). Its core belief is that "everything in the biosphere is interdependent, intrinsically valuable, and sacred" (Taylor 2010, 102). This is the belief system to which the founders of the Society for Conservation Biology ascribed. Furthering this belief is a more eco-centric and anti-consumerism philosophy laid out in four postulates (Lewis 1992). The first postulate is that decentralization of institutional power and control, leading to local autarky, is necessary for ecological and social health. The second postulate is that technological advancements, if not scientific progress itself, are inherently harmful and dehumanizing. The third postulate is that primal or primitive people exemplify how to live in harmony with nature (and with each other). Fourth, and finally, is the postulate that the capitalist market system is inescapably destructive and wasteful. The argument against these postulates is more practical than philosophical: that after several decades of preaching these philosophies, the public remains, as before, "wedded to consumer culture and creative comforts" (ibid.).

As technology advances, its proponents challenge these ethics as they are based on a premise of the division of humans from nature. For instance, Gruen, Jamieson, and Schlottmann propose that: "If we are part of nature, then everything we do is part of nature, and is natural in that primary sense. When we domesticate organisms and bring them into a state of dependence on us, this is simply an example of one species exerting a selection pressure on another" (2013, 232). Yet Jeremy Rifkin described biotechnology this way in 1983:

> as bioengineering technology winds its way through the many passageways of life, stripping one living thing after another of its identity, replacing the original creations with technologically designed replicas, the world gradually becomes a lonelier place. From a world teeming with life . . . we descend to a world stocked with living gadgets and devices.
>
> (Rifkin, cited in Lewis 1992, 123)

This question of humans manipulating nature either by direct means (gene drives), or in the normal co-evolution of species reflects upon the broader concepts of "wildness" and "wilderness." Gruen, Jamieson, and Schlottmann (2013) suggest that a place is wild when its order is created according to its own principles of organization and that what counts as wilderness is determined not by the absence of people, but by the relationship between people and place. Jack Turner offers the following view of wildness:

> To construct a new conservation ethic, we need first to understand why we impose a human order on non-human orders. We do so for gain, the gain being prediction, efficiency, and hence, control. Faced with the accelerating destruction of ecosystems and the extinction of species, we believe our only option lies in increased prediction, efficiency, and control. So we fight to preserve ecosystems and species, and we accept their diminished wildness. This wins the fight but loses the war, and in the process we simply stop talking about wildness.
>
> (cited in Gruen, Jamieson, and Schlottmann 2013, 204)

These strict interpretations of what is wild and what constitutes wildness denies a middle ground where responsible use and management might create a balanced, sustainable relationship.

The New Eco-Radicals?

In his book *Eco-Terrorism*, Liddick (2006, 78) describes the emergence of eco-radical groups: "As traditional methods for bringing about change fail, or do not bring change quickly enough, disaffected activists break off and form a new group or movement that advocates more extreme methods." The original environmental radicals challenged society to reorient itself. They asked, or demanded, that society "seek a steady-state relationship with all of nature's creations, wherein human attitudes and actions dominate no one and nothing" (Scarce 2006, preface). Or, as Sierra Club's first executive director and founder of Friends of the Earth David Brower says, "We cannot go on fiddling while the Earth's wild places burn in the fires of our undisciplined technology" (ibid., 20, citing Brower). Brower goes on to say that the environmental movement must "stir people up, not just do the expected" and that "[w]ith the end of nature ever closer, creativity in environmental battles is what's called for, not compromise" (ibid.). When asked what led them to adopt their "uncommon" principles, activists responded that it was not some clear, rational, deductive thought process; it emerged out of an ecological consciousness that came from the heart, not the head, which experienced the natural world (ibid.). This affect-guided ethics is discussed also by McCloskey, who states:

> I don't think the direct action techniques will work forever . . . but they will work for a while. That's part of pragmatism: to keep trying things, seeing what works. As long as it works, you pursue it. When it stops, you go on to something else.
>
> (ibid., citing McCloskey)

I agree with Liddick's description of the past and believe that the conservation community is again shifting toward a new "eco-radicalism," although this does not manifest in what were in the past sometimes violent, illegal activities (i.e., Earth First!). Emerging instead are groups that are proposing to change how conservation is practiced and funded. Such groups are using what some might perceive as "extreme methods" (i.e., gene drives and other biotechnologies). Three groups in particular—Revive & Restore, Island Conservation, and Conservation X Labs—have begun a conversation about how and why the conservation community should look to technology to solve pressing conservation needs.

In a recent press release, Island Conservation explains that:

> damaging, invasive alien rodents have invaded an estimated ninety percent of our world's island archipelagos. Eradicating this threat from islands provides significant public and conservation benefits. However, today's eradication tools limit the rate and scale at which this crisis can be curbed.
>
> (Island Conservation 2016)

Revive & Restore have been researching de-extinction and promoting the potential use of gene drives to combat avian malaria (Revive & Restore n.d.). Finally, Conservation X Labs is looking to disrupt the current conservation models by harnessing "exponential technologies, open innovation, and entrepreneurship to dramatically improve the efficacy, cost, speed, scale and sustainability of conservation efforts to end human induced extinction" (Conservation X Labs n.d.). While these methods may seem extreme to some, they present a solution to what others have called a "crisis discipline," implying that the current conservation methods are not working and there needs to be a paradigm shift in both thinking and pace.

So are we seeing a larger shift in the conservation community towards technological solutions? Possibly.

One of the first pieces of evidence of this shift came with the writings of Kent Redford and the first meeting of conservationists and synthetic biologists. In their seminal paper, "Synthetic Biology and Conservation of Nature: Wicked Problems and Wicked Solutions," Redford, Adams, and Mace (2013) described the reasons for bringing these two groups together. Redford and his co-authors stated:

> Conservation as a practice has frequently been backwards looking, focusing on reducing loss or on maintaining a status quo, an approach that has clearly not been effective in conserving biodiversity. Potential major shifts in the relationship between humans and nature such as those represented by synthetic biology would be better engaged with early and deeply. Yet of the hundreds of conservation practitioners with whom we have spoken, only a few had even heard of synthetic biology and had any sense of the changes it may bring. In order to expedite the engagement between the two fields, we have organized a meeting entitled "How will synthetic biology and conservation shape the future of nature?
>
> (Wildlife Conservation Society 2013)

Nonetheless, the conservation community has been reluctant to embrace emerging technologies (Redford, Adams, Carlson, and Mace 2014). During the first meeting of conservationists and synthetic biologists, Kent Redford made two notable observations. First, that conservation practice tends to be reactive to change driven by other fields of human endeavor. The techniques and approaches used have been honed by decades of experience, both trial and tribulations, and are well-defined, with established practices and procedures. The second observation was that attitudes toward innovation are closely linked to attitudes towards risk. Conservationists tend to be risk-averse in their practice of conservation. The stakes are high, the fear of failure constantly reinforced, and the priority is generally to minimize risks of irreversible consequence of their interventions, especially given many practitioners' experiences of the outcomes from experiments in conservation.

It is interesting to reflect back on Liddick and how he described the dynamic between the underground movement (eco-radicals, e.g., Earth First!) and the aboveground actors (traditional mainstream environmental organizations, e.g., Sierra Club) in the environmental movement. He suggests that underground activists remain anonymous and isolated, and that their success depends critically on aboveground members in the movement who provide support and direction:

> The maintenance of aboveground operatives is especially important in providing at least a perception of legitimacy to the given movement and communicating agendas to underground operatives; aboveground sector is essential if there is to be any hope for political change.
>
> (Liddick 2006, 70)

Whether or not mainstream conservation organizations publicly or privately support gene drives, however, is still under debate. At both the 2016 IUCN World Congress and the 2016 Convention on Biological Diversity's (CBD) bi-annual meeting, one could witness the emergence of the new eco-radicals. They were discussing how genetic technologies could be utilized for conservation purposes, while the mainstream conservation community, in the form of motions and statements, expressed reluctance. The mainstream community, however, did open the door slightly (see below). The political and moral will to save a particular species gave rise to some of the tactics used by "eco-radical" groups in the past. If gene drives follow a democratization path similar to other biotechnologies, leadership within the conservation community will be needed in order to discuss when, where, how and if they should ever be used.

Democratization of Gene Drives

In 2016, nearly 5,600 students (mostly under the age of 25) from 42 countries participated in the International Genetically Engineered Machines (iGEM) competition. One of those teams attempted (but failed) to build a gene drive system. While iGEM has a strict "no release" policy and has since then developed a policy pertaining to gene drives, the emergence of a team attempting to build a gene drive using standardized parts was alarming to many (Swetlitz 2016). However, it probably should not have been. Since 2010, the iGEM competition has rapidly expanded, along with the do-it-yourself biology (DIYbio) movement. Spurred by the convergence of economic and social forces, the pursuit of laboratory-based activities and access to biotechnology are no longer domains occupied exclusively by trained scientists working in academia, government, or industry. Low-cost technologies, access to funding and other reductions in barriers to entry have resulted in a broad range of people conducting sophisticated lab activities—including citizen scientists, hobbyists, and entrepreneurs. As gene editing techniques, and possibly gene drives, become more accessible and democratized, regulatory systems will need to re-engineer themselves to keep pace with the rapidly expanding ecosystem of actors. In fact, a new system of governance may now be warranted. In "Governance: Learn from DIY Biologists" (Kuiken 2016), I explored the democratization of advanced biotechnologies such as CRISPR-Cas9 and gene drives, to assess how the broader scientific community could adopt a more decentralized governance strategy. A similar international governance system may be needed to address the new eco-radicals that are developing gene drives for conservation purposes.

The equipment and reagents that are needed to use CRISPR-Cas9 are already readily available to DIY biologists. Members of the teams that participated in the 2015 iGEM competition—including high-school students and users of community labs around the world—received CRISPR-Cas9 plasmids in their starting kits. These kits contain more than 1,000 standard biological parts known as BioBricks, the DNA-based building blocks that participants need to engineer a biological system for entering into the competition. Other components of the CRISPR-Cas9 system are also available from the iGEM registry (iGEM n.d.).

One development that has increased anxiety about the widespread use of CRISPR-Cas9 is the work of synthetic biologist Josiah Zayner, founder of the Open Discovery Institute in Burlingame, California. Thirty days after launching his campaign on the crowdfunding website Indiegogo in November 2015, Zayner had raised almost 34,000 dollars to fund the production and distribution of DIY CRISPR kits—supposedly to help people "learn modern science by doing." He has since raised more than 62,000 dollars, six times his original goal. But the concern about Zayner's project arises not because it gives people outside conventional labs more capabilities than they would have otherwise. DIY biologists already use various tools to assemble DNA fragments in bacteria and yeast—the microorganisms that he supplies in his kits. Zayner's campaign is worrisome because it does not seem to comply with the DIYbio.org code of conduct, primarily the rule to "adopt safe practices" (DIYbio 2011). The video that accompanies his campaign zooms in on Petri dishes containing samples that are stored next to food in a refrigerator, which violates basic lab safety protocols. More than anything, Zayner's campaign is a reminder of the myriad ways in which researchers—conventional or otherwise—can now get their work funded. With the advent of crowdfunding platforms, researchers no longer need to go through the cumbersome and long process of traditional funding agencies, and can quickly have a project funded, sometimes within 30 days or less. The case of Josiah Zayner highlights the need for a more decentralized governance for everyone, including DIY biologists. Codes of conduct will be needed to establish appropriate norms for government funding and regulatory agencies, for people working both within and outside conventional research settings, for the directors of community labs, and for the developers of crowdfunding platforms.

Much of the concern surrounding the democratization of applications like gene drives is related to biosecurity and the fear that they could be used to harm. While the DIYbio community has established norms to reduce the risks of such occurrences, the overall issue of biosecurity and how these technologies are used deserves a robust public debate. The increasing investment by the United States military in these technologies raises similar questions: in terms of biosecurity, who should have access to these technologies and to the subsequent international treaties that govern them?

Militant Environmentalists

Between 2008 and 2014, the United States invested approximately 819 million dollars in synthetic biology research. The data presented in Figure 3 shows that federal funding for synthetic biology is increasing rapidly—from almost negligible amounts in 2005 to more than 200 million dollars in 2014. Since 2012, the majority of synthetic biology funding is coming from the Defense Advanced Research Projects Agency (DARPA). In 2014, nearly 60 percent of all funding in the United States—nearly 110 million dollars—came from DARPA (Kuiken 2015). While the phrase "synthetic biology" does not show up in DARPA budget documents until 2011, funding for "synthetic fuels," "synthetic cells," and "synthetic chromophores" began to appear in 2008 and continued through 2010.

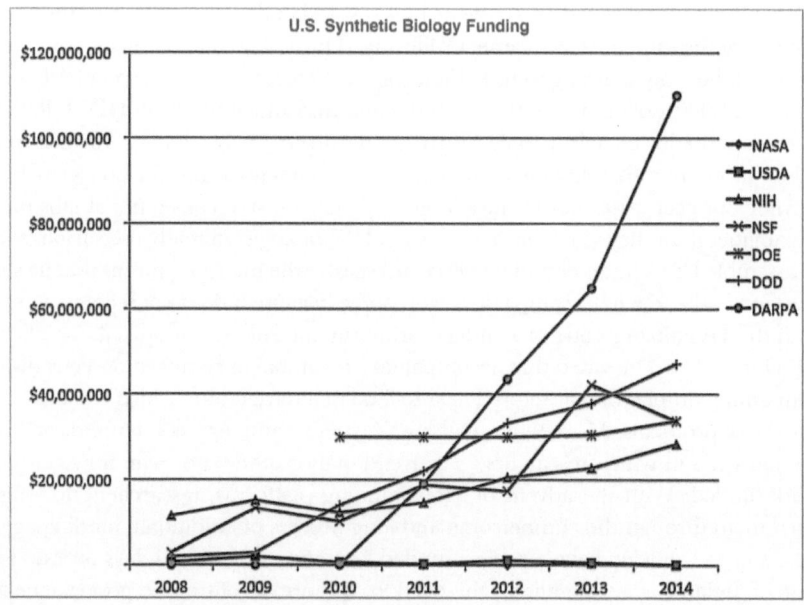

Figure 3 United States' federally funded research in synthetic biology.
Courtesy of The Synthetic Biology Project 2015.

Generally, award levels for Department of Defense (DOD)-funded research are classified. However, unclassified funding data was found for synthetic biology projects within five DOD programs, using the Defense Technical Information Center investment budget search database (DTIC 2016). DOD has significantly broadened and deepened its support for synthetic biology work, which now includes research conducted by the Army, Navy, and by the Office of Secretary of Defense, including the Chemical and Biological Defense Program, and DARPA (Kuiken 2015). DOD confirmed support for 18 projects in the Naval Biosciences program, but it did not provide funding amounts. Because the budgets for DOD projects are not publicly itemized, it is difficult to determine how much was allocated towards synthetic biology projects. To provide some context, the report approximated the total spending on synthetic biology-related projects to be 50 percent of the total budget listed (Kuiken 2015).

DARPA has developed five programs, listed here in chronological order, that demonstrate the growing complexity of the biotechnologies under development that could have a direct impact on the environment. The first program is called "Living Foundries," launched in 2013. This program:

> seeks to transform biology into an engineering practice by developing the tools, technologies, methodologies, and infrastructure to increase the speed

of the biological design-build-test-learn cycle while significantly decreasing the cost and expanding the complexity of systems that can be engineered. The technologies and infrastructure developed as part of this program are expected to enable the rapid and scalable development of transformative products and systems that are currently inaccessible. Examples include novel materials, industrial chemicals, pharmaceuticals, and improved agricultural products.

(DARPA 2016b)

The second program is called "Biological Robustness in Complex Settings" (BRICS). Launched August 2014, this program:

seeks to develop the fundamental understanding and component technologies needed to engineer biosystems that function reliably in changing environments. A long-term goal is to enable the safe transition of synthetic biological systems from well-defined laboratory environments into more complex settings where they can achieve greater biomedical, industrial, and strategic potential.

(DARPA 2016c)

The third program, named "Safe Genes," was launched in September 2016 and is all about gene drives. This program is meant to "create biological capabilities that enable the safe pursuit of advanced genome editing applications." According to the website:

Implementation of a 'safety first' approach to the development of next generation biotechnologies and genome editing tools and their derivative technologies (e.g., gene drives) will foster, and even accelerate, responsible innovation while mitigating the risk of unintended consequences. The Safe Genes program will provide new insights into what is possible, probable, and vulnerable with regard to genome editing biotechnologies and their derivative applications, create novel tools to enable predictable and reversible control of gene editors, and counter unwanted genome editing activity and outcomes.

(DARPA 2016d)

The program is designed to develop safety mechanisms for gene drives in order to advance their eventual use in the environment.

The fourth program, Insect Allies, was launched by DARPA in November 2016. This program:

Aims to transform certain insect pests into "Insect Allies," by modifying insects to disseminate targeted genetic payloads to plant populations in order to protect crops from potential plant pathogens that are either naturally occurring or are intentionally designed and released to cause harm.

(DARPA 2016e)

Finally, the fifth DARPA program, entitled "Ecological niche-preference engineer-ing," is being launched in 2017. This program centers on:

> the development of technologies that enable the genetic engineering of an organism's preference for a niche (e.g., temperature range, food source, and habitat). DARPA envisions creating genetic engineering strategies to control and alter the niche preferences of organisms to reduce economic, health, and resource burdens. A fundamental component of this work will be to expand our understanding of the genetic, epigenetic, and molecular contributors to the establishment of niche preference.
>
> (DARPA 2016f)

DARPA's interest in ecological manipulation has shifted demonstrably from organism- and species-level to ecosystem management more broadly.

Because the largest proportion of funding for advanced biotechnologies (includ-ing gene drives) now comes from the United States defense agencies, in practice, researchers are forced to apply for such funding. Take, for instance, the Safe Genes program, which promotes the use of biological technologies for environmental purposes. At the time of writing, there are no other federal research programs focusing on safety mechanism for gene drives. So if a research group or conserva-tion organization were interested in developing safety mechanisms for gene drives, their only option for funding would be to apply for defense spending. How is this type of funding perceived by the conservation community? Are we comfortable with the "militarization" of conservation? What are the potential military uses of these technologies and what could it mean in terms of international treaties?

With the advent of these programs, particularly inside DARPA, could one make an argument that DARPA and the United States military is becoming a "militant" environmentalist? I would argue that this is indeed the case, based on the pro-gression of programs that DARPA is developing in the environmental sphere. As noted above, these programs range from the development of manufacturing plat-forms to produce chemicals and materials to engineering entire ecosystems. They are defensive in terms of protecting the United States (and the environment) from potential biosecurity threats, but are also offensive in supplying the military with new capabilities to produce materials and chemicals. DARPA's interest in develop-ing safety mechanisms for gene drives suggest that they eventually would want to deploy them. For what purpose it is still unclear, but DARPA's mission is "to make pivotal investments in breakthrough technologies for national security" (DARPA 2016a) and not to save a particular species from extinction. This dichotomy raises the specter of the United States military as a "militant" environmentalist.

I am not suggesting that the United States military is developing offensive bio-logical weapons, but how the world perceives these programs and what they are being used for deserves attention, and could be debated in the Environmental Modification Convention (ENMOD), formally the Convention on the Prohibi-tion of Military or Any Other Hostile Use of Environmental Modification Tech-niques, and in biological and chemical weapons treaties more broadly.

Entered into force in 1978, the ENMOD Convention defines environmental modification techniques as:

> changing—through deliberate manipulation of natural processes—the dynamic, composition or structure of the earth, including its biota, lithosphere, hydro-sphere, and atmosphere, or of outer space. Changes in weather or climate patterns, in ocean currents, or in the state of the ozone layer or ionosphere, or an upset in the ecological balance of a region are some of the effects which might result from the use of environmental modification techniques.
>
> (ENMOD 1978)

Article 1 of the Convention states that:

> Each State Party to this Convention undertakes not to engage in military or any other hostile use of environmental modification techniques having widespread, long-lasting or severe effects as the means of destruction, damage or injury to any other State Party.
>
> (ibid.)

The term "widespread" is defined as "encompassing an area on the scale of several hundred square kilometers"; the term "long-lasting" is defined as "lasting for a period of months, or approximately a season; and "severe" is defined as "involving serious or significant disruption or harm to human life, natural or economic resources or assets."

The Convention does allow for peaceful uses of environmental modification techniques and the possible exchange of scientific and technological information between parties. It also establishes a Consultative Committee of Experts, which would meet on an ad hoc basis when so requested by a party, in order to clarify the nature of activities suspected to be in violation of the Convention. While this Convention was developed in response to manipulating weather patterns during the Vietnam War, its aims and definitions seem to capture the manipulations inherent in gene drives. Hence, a nation state or a concerned party could invoke this treaty if they perceived a nation state to be in violation. This is not an unlikely scenario, in light of the scale of research currently being developed inside DARPA. Such issues are crucial when attempting to conduct a robust public (global) dialogue about when, where, and if one entity chooses to use gene drives.

Trust, Acceptance, Justice

Conservation has been described as a "crisis discipline" (Johnson et al. 2016), raising questions such as "are we willing to risk the global loss of a species as a result of unintended dispersal of modified individuals back to their native range or benefit from the control efficiencies that CRISPR-Cas9 gene drive technology could offer?" (Webber et al. 2015). Scenarios like these are used as prompts for why novel technological solutions, like gene drives, should be explored to solve conservation problems. Fortini et al. (2015) explore one such example of the use of genetic technologies to solve a conservation issue in their article entitled "Large-Scale Range Collapse of Hawaiian Forest Birds under Climate Change and the Need for 21st Century Conservation Options." The authors state that:

> unless conservation managers are willing to risk future range collapse and extinction of species, a concerted effort to explore the viability of novel long-term solutions that decouple projected climate shifts from species declines must be started before it is too late to successfully implement them.
>
> (ibid., 13)

They go on to say that "the longer it takes to fully develop novel long-term solutions, the greater the amount of conservation resources that will need to be spent on buying-time actions" (ibid., 15). This, of course, leaves fewer resources to develop long-term solutions under the current funding scenarios. They suggest further that the regulatory frameworks and public support for such novel solutions need to be addressed now, as they can be a significant challenge to implementing any novel conservation strategy.

Trust is a critical component in any negotiation, particularly one involving new technologies and environmental protection. Trust-based environmental regulation is defined as a specific regulatory style that involves openness and cooperation in interaction between regulated, regulators, and third-party stakeholders in order to achieve environmental objectives (Lange and Gouldson 2010). A potential framework described by Lange and Gouldson, for which gene drives could present a novel case study,

> is to open up new ways for participants in regulatory regimes to engage in collective action, to go beyond a perception of regulation driven by the competing interests of individual actors, and thus, to open up new channels of influence for behavioral change towards greater environmental protection.
>
> (ibid., 5235)

This system bodes well for dealing with commons issues, under which category gene drives clearly fall, and could be used to support Esvelt's argument in Chapter 1 (this volume) that gene drives should only be developed locally.

When dealing with issues of the commons there is a tipping point of acceptance that can occur (Lacey and Lamont 2014). For example, any gene drive application will need multiple release sites in order for the drive to spread properly. Therefore, multiple agreements will need to be acquired from potential landholders and governments (local, state, country) for field trials to move forward. The decision of a single stakeholder can affect other stakeholders who may be interconnected, or are part of the same ecosystem in which the gene drive is proposed to be released. However, if a critical mass of stakeholders agrees to the release, there may be little point for those opposing gene drives to continue their objection, because the actions of those who agreed makes it so there is no longer any way for detractors to avoid the perceived risk (ibid.). Take for example the debate in Key Haven, Florida, where the local mosquito control district is contemplating utilizing genetically engineered mosquitoes. A map of Key Haven demonstrates the issue of how one landholder's decision could affect another in relation to an application like gene drives. Regardless of whether or not an individual land-holder rejects the release, if, and when, it is approved they would not be able to keep the mosquitoes off their property. Extrapolating this scenario out, the same could potentially hold true for countries bordering other countries that implement a gene drive application.

The citizens of Key Haven have intensely debated whether or not to allow such a field trial (Klinger 2016). Polls suggest growing acceptance of GM mos-quitoes: 51 percent of respondents had a positive view toward GM mosquitoes in 2013 (Widmar and Tyner 2016), which has grown to 78 percent in 2016 (Eise 2016). This increase in support could be attributed to the rise of Zika in the news. Although the FDA granted approval to the company to release the genetically modified mosquitoes, the Florida Keys Mosquito Control District decided to allow the citizens to vote on whether to approve the release (Axford 2016). Only a small majority of voters across the county supported the use of the technology (57 per-cent), and in the proposed field trial site, a strong majority opposed it (65 percent), representing a classic example of "not-in-my-backyard" (Servick 2016).

Confusion regarding which agencies are responsible (if any) for gene drives can lead to public mistrust, as discussed in Chapter 1. A clear example of this con-fusion surrounds the regulatory review of Oxitec's Diamondback Moth (Oxitec 2016a) and mosquito (Oxitec 2016b). While neither of these are gene drives, both utilize similar genetic technologies in order to reduce the population size in the wild. Yet because the moth is considered a plant pest, its application is reviewed by the United States Department of Agriculture, while the mosquito is being reviewed by the United States Food and Drug Administration, because the genetic technologies used to crash the population are considered an animal drug. Both have been approved for field trials. Yet, as of 2016, neither has moved forward. In my opinion, the reason for this is public reluctance and distrust.

International Treaties and Obligations

Gene drives are designed to move. In fact, they don't work unless they spread throughout the entire species and its habitat range. Therefore, gene drives are not going to respect a country's border. As such, it will be nearly impossible to acquire acceptance from all affected parties and based on its global impact, which creates the potential to "deny humanity and all other residents of this earth the respect that they are owed" (Hale 2013, 217). Therefore, should gene drives and other genetic technologies designed to alter, manage, and control ecosystems be discussed in the same context as geoengineering? Geoengineering has typically been associated with climate control, but the bases of its applications are similar to gene drives: large-scale manipulations of natural systems and processes. Decisions associated with geoengineering, ecosystem management, disease vector control, or other large-scale biological manipulation have impacts beyond their initial release point. Those subsequent impacts and the publics they affect should be considered in an international context. Because the properties of gene drives make it likely that the release in one country may impact another, it raises the need for international agreements, and the Convention on Biological Diversity (CBD) seems to be a logical starting point.

In December 2016, during the Thirteenth Meeting of Conference of the Parties (COP) to the Convention on Biological Diversity, gene drives were discussed under the auspices of synthetic biology. Since 2010, the discussion has focused on whether synthetic biology should be classified as a new and emerging issue. Defining synthetic biology as such would officially state that it "needs urgent attention by the Subsidiary Body on Scientific, Technical and Technological Advice (including how it impacts biodiversity)" (Convention on Biological Diversity 2003). This action would enable the CBD to develop new guidance on how synthetic biology and its applications could be utilized in the future by a member state.

During the COP in 2016, there was intense debate about gene drives, with multiple side-events addressing their benefits and risks. There was also formal debate within the working group on synthetic biology on whether a moratorium should be placed on the release of gene drives and on any form of research using this application. Three proposals were under discussion inside the working group for synthetic biology. The proposals ranged from a moratorium on both research and release of gene drives to restating a precautionary approach toward gene drives (using the same language that is used for synthetic biology). Eventually, the final, agreed-upon text that was submitted to the Secretariat "*reiterates* paragraph 3 of decision XII/24 and *notes* that it can also apply to some living modified organisms containing gene drives" (Convention on Biological Diversity 2016a; 2016b). Decision X11/24 paragraph 3, which was adopted at the COP 12 in 2014, "*Urges* Parties and *invites* other Governments to take a precautionary approach, in accordance with paragraph 4 of decision XI/11." This issue will be taken up by an open-ended online forum and by the Ad-Hoc Technical Expert Group in 2017. These groups will make recommendations to the Secretariat for further negotiations at the 14th COP to be held in 2018.

Similar deliberations pertaining to gene drives have been occurring at the International Union for Conservation of Nature (IUCN). The IUCN is the world's leading association of conservation agencies, including nation states, government agencies, and non-governmental organizations. It is the only environmental organization with observer status at the United Nations General Assembly. The IUCN has been involved in the CBD since its initial drafting and throughout its development. At their World Conservation Congress in 2016, there were similar side-events to the ones held at the CBD, along with debates on moratoriums for gene drives. Finally, the IUCN members voted on a motion, which in part calls upon the Director General of the IUCN to undertake an assessment of both synthetic biology and gene drives. However, the IUCN motion went further than the final text of the CBD by stating that the IUCN will refrain "from supporting or endorsing research, including field trials, into the use of gene drives for conservation or other purposes until this assessment has been undertaken" (IUCN 2016).

Final Thoughts

The relationship between technology and conservation is a complicated one. Does technology's use resign us to the notion that we failed, and that consumerism, environmental exploitation, and unsustainable population growth are taking precedence over traditional conservation? Or could they be viewed as part of the collective evolution of human-kind and nature? These ethical considerations are discussed in detail in Chapter 2 (this volume). Understanding and evaluating the ecological effects of these genetic technologies will require broad interdisciplinary convergence and an enhanced ability on the part of governmental systems to adapt to the rapidly changing technologies and to the new actors who will have access to them.

Earth First! was formed in 1980, representing the first radical environmental organization devoted to civil disobedience and sabotage as a means of environmental resistance (Taylor 2010, 74). Over time, traditional environmental advocacy groups, including the Wilderness Society and Sierra Club, gave way to more radical groups like Greenpeace, Sea Shepherd Conservation Society, and eventually the Earth Liberation Front. I am not suggesting that groups exploring gene drives for conservation are similar in terms of the tactics and what are often the illegal activities undertaken by eco-radical groups; I am instead suggesting that the underlying philosophies and circumstances that enabled the emergence of such groups may be reemerging in the context of using genomic technologies for conservation. We are arguably at a tipping point, not just in terms of our ability to conserve and protect nature, but also in terms of the direction in which the environmental community is moving or not moving. Will the new eco-radicals change the hearts and minds of the environmental community? Will the environmental community embrace DARPA as a partner in conservation? And would such an endorsement, if it occurs, change the very identity of the environmental community and how it resolves the conservation problems of today and tomorrow?

From the ethos of a new environmental conservation strategy to the rapid democratization of genetic technologies, and finally to the increasing investments from the military—when we talk about the democratization of technology, technology transfer, or even economic fairness and justice, what do we really mean? If we are serious about these issues, we must accept that people may choose to use these technologies in a way to which we might not all agree. In other words, one person's vigilante could be another person's savior.

References

Axford, William. 2016. "Key Haven Residents to Vote on Genetically Modified Mosquitoes in Nonbinding Referendum." *FL Keys News*. April 23. Available at: www.keysnet. com/2016/04/23/508107/key-haven-residents-to-vote-on.html

Conservation X Labs. n.d. "Exponential Conservation." Available at: http://conservationxlabs. com/home

Convention on Biological Diversity. 2003. "COP 9 Decision IX/29." Available at: https:// www.cbd.int/decision/cop/default.shtml?id=11672

————. 2016a. *Draft Decision Submitted by the Chair of Working Group II.* December 13. Available at: https://www.cbd.int/doc/c/1606/8ba5/0856c9f00b7a960f7057be65/cop-13-wg-02-crp-22-en.doc

————. 2016b. *Draft Decision Submitted by the Chair of the Working Group II.* December 16. Available at: https://www.cbd.int/doc/c/816f/ffca/103ebb49656424c0858f129a/cop-13-l-34-en.doc

Cramer, Phillip F. 1998. *Deep Environmental Politics: The Role of Radical Environmentalism in Crafting American Environmental Policy.* Santa Barbara, CA: Praeger.

DARPA. 2016a. Defense Advanced Research Project Agency. Available at: www.darpa.mil/about-us/about-darpa

————. 2016b. "Living Foundries Program." Available at: www.darpa.mil/program/living-foundries

————. 2016c. "Biological Robustness in Complex Settings (BRICS)." Available at: www.darpa.mil/program/biological-robustness-in-complex-settings

————. 2016d. "Safe Genes Proposers Day." Available at: www.darpa.mil/news-events/safe-genes-proposers-day

————. 2016e. "Insect Allies Program." Available at: https://www.fbo.gov/utils/view?id=40638c9e7d45ed8310f9d4f4671b4a7b

————. 2016f. "Ecological Niche Preference." Young Faculty Award: DARPA-RA-16-63. Available at: https://www.fbo.gov/utils/view?id=c1540b48aa08624b27f4dc5e7cdf94fe

DIYbio. 2011. *Draft DIYbio Code of Ethics from North American Congress.* Available at: https://diybio.org/codes/code-of-ethics-north-america-congress-2011

DTIC. 2016. Defense Technical Information Center. Available at: www.dtic.mil/dodinvestment/#/home

Eise, Jessica. 2016. "Public Supports Use of GMO Mosquitoes to Fight Zika Virus." *Purdue University: Agriculture News.* February 22. Available at: https://www.purdue.edu/newsroom/releases/2016/Q1/survey-public-supports-use-of-gmo-mosquitoes-to-fight-zika-virus.html

ENMOD. 1978. "Convention on the Prohibition of Military or Any Other Hostile Use of Environmental Modification Techniques." Geneva, Switzerland.

Fortini, Lucas B. et al. 2015. "Large-Scale Range Collapse of Hawaiian Forest Birds under Climate Change and the Need for 21st Century Conservation Options." *PLOS One* 10: e0144311.

Gruen, Lori, Dale Jamieson and Christopher Schlottmann. 2013. *Reflecting on Nature: Readings in Environmental Ethics and Philosophy.* New York: Oxford University Press.

Hale, Benjamin. 2013. "Remediation vs. Steering: An Act-Description Approach to Approving and Funding Geoengineering Research." In *Designer Biology: The Ethics of Intensively Engineering Biological and Ecological Systems.* Edited by John Basl and Ronald Sandler, 195–218. Lanham, MD: Lexington Books.

iGEM. n.d. *International Genetically Engineered Machines Competition Registry.* Available at: http://parts.igem.org/CRISPR

Island Conservation. 2016. "Investigating Suitability of Genetic Biocontrol of Invasive Rodents on Islands." September 4. Available at: https://www.islandconservation.org/release-investigating-suitability-genetic-biocontrol-invasive-rodents-islands

IUCN. 2016. "Development of IUCN Policy on Biodiversity Conservation and Synthetic Biology." IUCN World Conservation Congress. Available at: https://portals.iucn.org/congress/motion/095

Johnson, Jeff et al. 2016. "Is There a Future for Genome-Editing Technologies in Conservation?" *Animal Conservation* 19: 97–101.

Klinger, Nancy. 2016. "Key Haven Residents Oppose GMO Mosquito Test." *WLRN News,* April 11. Available at: http://wlrn.org/post/key-haven-residents-oppose-gmo-mosquito-test

Kuiken, Todd. 2015. *U.S. Trends in Synthetic Biology Research Funding.* Washington, D.C.: Woodrow Wilson Center. Available at: www.synbioproject.org/publications/u.s-trends-in-synthetic-biology-research-funding

———. 2016. "Governance: Learn from DIY Biologists." *Nature* 531: 167–168.

Lacey, Justine and Julian Lamont. 2014. "Using Social Contract to Inform Social Licence to Operate: An Application in The Australian Coal Seam Industry." *Journal of Cleaner Production* 84: 831–839.

Lange, Bettina and Andy Gouldson. 2010. "Trust-Based Environmental Regulation." *Science of the Total Environment* 408: 5236–5243.

Lewis, Martin W. 1992. *Green Delusions: An Environmental Critique of Radical Environmentalism.* Durham, NC: Duke University Press.

Liddick, Donald. 2006. *Eco-Terrorism: Radical Environmental and Animal Liberation Movements.* Santa Barbara, CA: Praeger.

Oxitec. 2016a. "Florida Keys Project." Available at: www.oxitec.com/health/florida-keys-project

———. 2016b. "Diamondback Moth." Available at: www.oxitec.com/agriculture/our-products/diamond-back-moth

Pepper, David. 1993. *Eco-Socialism: From Deep Ecology to Social Justice.* New York: Routledge.

Redford, Kent, William Adams, Rob Carlson, and Georgina Mace. 2014. "Synthetic Biology and the Conservation of Biodiversity." *Oryx* 48: 330–336.

Redford, Kent, William Adams, and Georgina Mace. 2013. "Synthetic Biology and Conservation of Nature: Wicked Problems and Wicked Solutions." *PLOS Biology* 11: e1001530.

Revive & Restore. n.d. Available at: http://reviverestore.org

Scarce, Rik. 2006. *Eco-Warriors: Understanding the Radical Environmental Movement.* Walnut Creek, CA: Left Coast Press.

Servick, Kelly. 2016. "Update: Florida Voters Split on Releasing GM Mosquitoes." *Science.* DOI: 10.1126/science.aal0350. Available at: www.sciencemag.org/news/2016/11/florida-voters-weigh-gm-mosquito-releases-what-are-issues

Swetlitz, Ike. 2016. "College Students Try to Hack a Gene Drive—And Set a Science Fair Abuzz." *Stat.* December 14. Available at: https://www.statnews.com/2016/12/14/gene-drive-students-igem

Synthetic Biology Project. 2015. *U.S. Trends in Synthetic Biology Research Funding.* Washington, D.C.: Woodrow Wilson Center. Available at: http://synbioproject.org/publications/u.s-trends-in-synthetic-biology-research-funding

Taylor, Bron. 2010. *Dark Green Religion: Nature Spirituality and the Planetary Future.* Berkeley, CA: University of California Press.

Webber, Bruce L. et al. 2015. "Opinion: Is CRISPR-Based Gene Drive a Biocontrol Silver Bullet or a Global Conservation Threat?" *Proceedings of the National Academy of Sciences* 112: 10565–10567.

Widmar, Nicole and Wally Tyner. 2016. "Public Supports Use of GMO Mosquitoes to Fight Zika Virus." Purdue University. Available at: "https://protect-us.mimecast.com/s/O578BaFw9lOViG?domain=extension.purdue.edu" https://extension.purdue.edu/Pages/article.aspx?intItemID=14074

Wildlife Conservation Society. 2013. *How Will Synthetic Biology and Conservation Shape the Future of Nature?* Available at: http://e.wcs.org/pdf/Synthetic_Biology_and_Conservation_Framing_Paper.pdf

Chapter 6

Controlling Our "Nature"

Gene Editing in Law and in the Arts

Lori Andrews

Scientists, venture capitalists, and breathless journalists enthusiastically tout the potential of CRISPR as a cure for everything from hunger to Zika to stupidity. But I've been involved in science policy long enough to be skeptical. After all, fetal tissue transplantation was supposed to cure Parkinson's disease. It didn't. Artificial chromosomes and stem cells were going to revolutionize medicine. They haven't. And—my personal favorite—an NIH scientist in the early 1980s said gene therapy would cure all genetic diseases by 1984.

Our collective fear of illness and death is so strong that as each new scientific intervention comes along, we are ready to run roughshod over important cultural values (such as access and quality) to implement that technology. As Steve Jones, an emeritus professor of human genetics at University College London, pointed out in a review in *The Lancet* of a book about gene editing and extinct species, "the four letters of the genetic code too often mutate from A, G, C, and T, to H, Y, P, and E" (2015, 125).

Currently, it is the CRISPR gene-editing technique that is causing a stir in the scientific community (Sander and Joung 2004). CRISPR is being extoled as a way to alter crops (Rodriguez 2016), revive extinct animals (The Long Now Foundation n.d.), create pets by miniaturizing pigs (Cyranoski 2015), and modify trees to grow more rapidly in order to improve reforestation efforts (Zhou et al. 2015). The medical applications of CRISPR are also being explored. In June 2016, a National Institutes of Health's advisory committee approved the first clinical trial of CRISPR in cancer therapy in the United States (Reardon 2016). In October 2016, a Chinese team tested CRISPR gene editing in a human for the first time (Cyranoski 2016).

As CRISPR enters our lives, what is the proper lens through which to assess and possibly regulate this next new thing in science? Profound decisions about shaping life should not be left to scientists alone. In Canada, a Royal Commission was chartered to recommend policies governing genetic and reproductive technologies as a whole (Royal Commission on New Reproductive Technologies 1993). The Commission used a variety of innovative methods to address these issues. They instituted an 800-number phone line so that citizens could detail their personal experiences with these technologies and express general opinions. In order to assess the values that defined Canadian life, they sought research and analysis from representatives of 70 disciplines on topics such as the psychological and social impacts of infertility, assisted reproduction, human zygote research, genetic testing, and the use of fetal tissue. The Commission determined that Canadian social values stressed non-commodification and non-objectification, as well as protection of the vulnerable. This led to the Commission's recommendation of bans on human cloning, paid surrogate motherhood, genetic enhancement, and sex selection for non-medical purposes.

In the United States, the Human Genome Project established the federal Ethics, Legal, and Social Implications (ELSI) program, and provided grants totaling nearly 5 percent of its budget to economists, philosophers, anthropologists, law professors, sociologists, and bioethicists to begin to anticipate the biopolitical issues that would arise with genetics. Their analyses led to enhanced community participation in research decisions, greater informed consent protection, and laws against genetic discrimination.

In this chapter, I add another set of voices to the discussion—that of novelists and artists. These creative observers have surfaced public concerns and dreams for technologies, while also addressing the individual and social impacts. Novelists wrote about *in vitro* fertilization, the internet, the atomic bomb, and many other technologies before they were developed. When Cleve Cartmill published the short story "Deadline" in the March 1944 edition of *Astounding*, Allied counterintelligence operatives immediately jumped to the conclusion that someone had leaked information to him about the secret project to develop the atom bomb. He had merely extrapolated from the existing scientific articles at the time.

Novelists have provoked legal policies with their works. Upton Sinclair spent the year 1904 working incognito in the Chicago meat-packing industry. His novel, *The Jungle* (1906), was a vivid description the perils of the slaughterhouses and meat-packing plants and a dissection of a corrupt political system. The book led to the passage of both the Pure Food & Drug and the Meat Inspection Acts during the Roosevelt Administration.

The visual arts, too, have provided social commentary and spurred political action. Following Dorothea Lange's publication of one of her photos, "Migrant Mother," in the *San Francisco News* in 1936, the U.S. government sent 10 tons of food to migrant workers (Meltzer 2000, 133), and ordinary people who saw the photo donated a total of 200,000 dollars to help establish a migrant camp for homeless workers (Gordon 2009, 217).

Both Sinclair and Lange were commenting on existing social problems. But in life science art, the artists additionally focus on the future—on the possible scientific manipulation of humans and other species. By engaging with artists using genetic technologies, we can begin to understand the social challenges and the biopolitical issues raised by CRISPR.

The History of Life Science Art

Life science art has evolved alongside life science itself. As far back as 1933, Alexander Fleming exhibited "germ paintings," which he sketched using bacteria on culture-medium-soaked papers (Stracey 2009, 496–497). Around the same time, the Museum of Modern Art in New York exhibited plants that were bred especially for aesthetic qualities. In 1936, the museum mounted an exhibition of delphiniums bred by the photographer Edward Steichen (Gedrim 1993). He created them by soaking seeds in a toxin that induced polyploidy (Stracey 2009, 496). The Nazi eugenics movement—the attempt to breed people like plants or animals—put a stop to museums' interest in the use of genetic concepts by artists. Museums did not want to glorify selective breeding or to appear to be endorsing anything like eugenics.

In recent years, however, there has been a resurgence of life science art. Oron Catts and Ionat Zurr saw the iconic 1997 biotech photo of a mouse with a human-shaped ear growing on its back and said to each other, "That's sculpture" (Andrews 2007b, 68–69). They talked their way into a residency at the Massachusetts Institute of Technology lab of Joseph Vacanti, who, with his brother Charles, had created the ear on the mouse by seeding bovine cartilage cells on biodegradable scaffolding. Catts and Zurr learned the tissue scaffolding technique and began their *Semi-Living Worry Dolls* project, taking inspiration from the Guatemalan tradition of giving small dolls to children at bedtime to which the children tell their worries. In 2000, Catts and Zurr seeded live cells on biodegradable scaffolding to create worry dolls and display them in a bioreactor in a gallery. Viewers used a microphone to whisper their worries about corporate biotechnology to the dolls (Figure 4).

Biopolitical Insights Raised by Artists

Life science artists predict, reflect, and influence the public's concerns and expectations about genetic technologies, sometimes before the technologies themselves are adopted into society. Their work has brought to the fore questions about the social and legal implications of genetic technologies. The following questions, raised and dealt with in myriad ways by life science artists, can help provide a starting point for analyzing CRISPR:

How can the public be involved in life science decisions?
How valid are the justifications for the technology?
How can the technology be designed to meet actual needs?
How do we collect data about the unintended consequences of the technology?
Who owns genetic materials, genetic processes, and genetic information?
How can we develop genetic technology with an ethics of caring and responsibility?

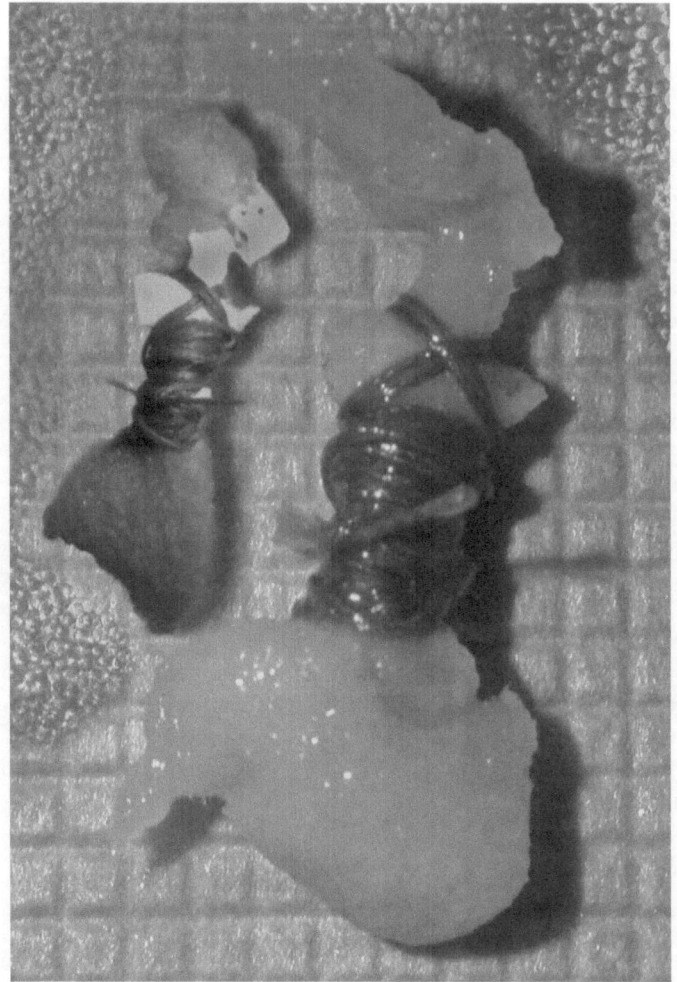

Figure 4 A Semi-Living Worry Doll C. McCoy Cell line, biodegradable/bioabsorbable poly-
mers and surgical sutures.

Courtesy of The Tissue Culture and Art Project (Oron Catts and Ionat Zurr), Ars Electronica 2000.

How Can the Public be Involved in
Life Science Decisions?

Artists use their work to critically assess genetic technologies or criticize the
manner in which they are being integrated into society. At SymbioticA, Oron
Catts and Ionat Zurr grew living pig tissue in the shape of wings. *If Pigs Could
Fly* demonstrated the possibilities of biotechnology, but also highlighted its limits.
One of the goals of the exhibit was to show people that their expectations about

biotechnology are excessive. People came into the gallery expecting to see pigs that could fly—instead, they saw tiny sculptures of tissue.

Artist Natalie Jeremijenko's work *One Trees* also combats genetic hype. She made 1,000 clones of a single tree. By planting these clones in different soil conditions and climates, she showed that, though genetically identical, they grew to look very different (Stracey 2009, 500; Jeremijenko 2007). Even the clones she exhibited in a single gallery were each unique—due in part to their different distances from light sources. The artist demonstrated that genes are not determinative and that the environment matters as well.

Artists understand that genetic technologies such as CRISPR will be adopted into a social context and that members of the public need to have a say in the development and implementation of biotechnology. For example, unlike Europe, the United States does not require the food industry to disclose when it uses genetically altered organisms. In its *Free Range Grain* exhibit, the Critical Art Ensemble (CAE) invites exhibit attendees to bring in food from their own homes to be tested. If the food is found to be a genetically altered product, the CAE leaves it to the visitors to "eat the food at their own risk." Their approach highlights the lack of information available to the public.

Similar artistic endeavors might be used with respect to CRISPR. In April 2016, the United States Department of Agriculture stated that a CRISPR-edited mushroom did not have to receive the agency's approval before its release to the market because genetic material (in this case, related to browning) had been deleted, not added (Waltz 2016). The agency's reasoning was that because the mushroom did not have inserted genes from "plant pest sequences," it did not fall under the regulations that apply to genetically engineered living entities (Friko 2016). The agency's decision about the mushroom prompted a scientist to state: "I am confident we'll see more gene-edited crops falling outside of regulatory authority" (Waltz 2016). CRISPR modifications have been proposed for soybeans (Jacobs et al. 2015), oranges (Jia and Wang 2014), potatoes (Wang et al. 2015), rice (Mikami, Toki, and Endo 2015), and wheat (Wang et al. 2014).

Yet can we really be sure that the deleting of genetic material is inconsequential? Shouldn't some testing be required about the impact of CRISPR-modified crops on humans before federal approval? And would the regulatory logic that deleting genetic material is perfectly acceptable also apply to humans? Many genetic diseases in humans are caused by small deletions or single nucleotide changes.

CRISPR is being treated by many scientists and the media as safe, precise, and effective, but the reality is far different. Since CRISPR gene editing targets particular small sequences, cuts can occur at identical sequences occurring elsewhere in the genome rather than at the intended spot (Rodriguez 2016). As a recent report of the National Academies of Sciences, Engineering, and Medicine points out, "proof-of-concept in a few laboratory studies is not sufficient in of itself to support a decision to release gene-drive modified organisms into the environment" (National Academies of Sciences, Engineering, and Medicine 2016, 14).

How Valid are the Justifications for the Technology?

Man has long attempted to control nature. In his work *Genesis*, artist Eduardo Kac played with that theme. Kac started with a single sentence in the Biblical chapter of Genesis: "Let man have dominion over the fish of the sea, and over the fowl of the air, and over every living thing that moves upon the earth." He then translated the Biblical sentence into Morse code, which gave him four characters—dots, dashes, letter spaces, and word spaces. He translated those four characters into the genetic code, providing a recipe for a new gene. Kac arranged for a biotechnology company to make the gene, which was inserted into bacteria in a Petri dish. When the creation was exhibited in a gallery, people around the world were able to become co-creators; they could log onto the internet and, with a click of their mouse, flash ultraviolet light into the gallery, causing the new gene to propagate and mutate (Kac n.d.; Kac 2005, 249–263). In another project, *GFP Bunny*, Kac arranged for the green fluorescent protein gene from a jellyfish to be inserted in a rabbit embryo; the resulting bunny glowed green under certain light (Kac 2005, 264–284).

Eduardo Kac has employed his own DNA in the service of art. The central piece of Kac's *Natural History of the Enigma* series is *Edunia*, a genetically engineered flower that is a hybrid of a petunia and Kac himself. Kac isolated and sequenced the gene from his own blood, specifically from a protein-coding gene sequence of his Immunoglobulin (IgG) light chain (Gambino 2013). Kac's gene produces a protein only in the red veins on Edunia's light pink petals, creating the image of human blood flowing through the veins of Edunia (Figure 5).

Artists and scientists both justify their work in synthesizing and modifying genes by referring to the long-standing practice of animal and plant breeding. In discussing his *GFP Bunny* project, Kac notes that while the project "may seem completely unprecedented, human direct influence on dog evolution goes back at least 15,000 years" (Kac 1998). Similarly, veterinarian and conservationist Alex Travis analogizes CRISPR "repairs" of a dog's genome to the long-standing practice of selective breeding (Chapter 8, this volume).

Just because a practice is long-standing does not mean it is appropriate. Slavery was long-standing, as was the law's refusal to let women hold property. The process of treating DNA like a set of Lego® can likewise be subject to challenge. Moreover, the process used by Kac or proposed by Travis is not analogous to selective breeding. Scientist Jon Gordon points out that there are enormous differences between selective breeding and a single gene being introduced or modified in a complex organism. Gordon notes that unlike selective breeding, where favorable alleles at *all* loci can be selected at one time, the current techniques try to improve a trait by affecting only *one* locus in isolation (1994). This single-gene approach has produced disastrous results. When a gene sequence shown to induce muscle hypertrophy in mice was inserted into a calf embryo, the animal did exhibit the desired trait initially, but later exhibited muscle deterioration (ibid.). The animal had to be shot. In a separate experiment, researchers genetically enhanced the

Figure 5 Eduardo Kac, Natural History of the Enigma. A transgenic flower with the artist's own DNA expressed in the red veins, 2003/2008.

Courtesy of the artist.

wings of flies to be 300 percent stronger than average. Instead of creating a super-fly, however, these flies couldn't even get off the ground because they were no longer able to move their wings fast enough.

In another study, researchers genetically enhanced mice to overexpress NMDA receptor 2B (NR2B), linked to long-term memory and increased cognitive and mental abilities. The resulting animals (called "Doogie [Howser]" mice) seemed to move more quickly through mazes than the mice that had not been altered (Tang et al. 1999). Immediately the question arose about whether such interventions should be undertaken on humans. Yet subsequent research by other scientists found the genetic intervention had a downside. The Doogie mice were more susceptible to long-term pain (Wei et al. 2001).

In addition, human's selective breeding of animals has caused direct harm to the animals; purebred dogs, for example, "are more likely to have disorders including cataracts, hypothyroidism, and a heart condition called dilated cardiomyopathy" (Wunderman 2016). The previous practice of selective breeding thus does not provide a mandate for CRISPR, but rather a caution against it.

How Can the Technology be Designed to Meet Actual Needs?

Life science artists have ridiculed biotechnologies designed to develop consumer products that do little to solve current problems. In Laura Stein's 1996 series, *Animal-Vegetable*, the artist placed animal-shaped molds over baby vegetables to coerce the vegetables to manifest abnormal shapes or facial expressions (Stein n.d.).

But life science art can also provide examples of how technology can be used to meet actual needs. Artist Mel Chin has been working with scientist Rufus Chaney since 1990 on *Revival Field*, the Lazarus of landfills (Chin n.d.; Strickland 2014). Through this biological and agricultural sculpture, Chin and Chaney have managed the ecological restoration of a toxic landfill located in St. Paul, Minnesota, by using a type of plant known as a hyperaccumulator. This type of plant extracts and retains heavy metals such as zinc and cadmium from soil, which can be reused once the plants are harvested and turned to ash (Youngs n.d.; Strickland 2014). Chin was "struck by the poetic nature of the project" and "conceived a sculpture in which plants and biotechnology would replace chisels and marble" (Comis 1995, 9). He was motivated to collaborate with Chaney, who had already been doing research in hyperaccumulators, by a feeling of "responsibility to the scientific advancements that could make this change possible" (ibid.).

In 2016, artist Orkan Telhan created a suitcase-sized biolab that sequences genes and also changes genomes via CRISPR. As artists begin to use CRISPR in their work, they will reflect what we do and do not need from CRISPR. In crime-ridden Chicago, perhaps artists will utilize the approach of Jalila Essaïdi. In *2.6 g 329 m/s*, she took spider silk produced in the milk of transgenic goats and seeded it with human skin cells to create bulletproof skin (Essaïdi et al. 2012).

CRISPR has also been referred to as a possible way of stopping the spread of the Zika virus. The bloodsucking female mosquitoes are the ones that bite and hence spread the Zika virus, while the nectar-feeding male mosquitoes do not. CRISPR may be able to insert "Nix," a gene that causes male genitalia to be introduced into female mosquitoes' genomes, which would cause virtually all offspring the female mosquito produced to be male (Krisch 2016). The number of female mosquitoes—and therefore the number of mosquitoes that can infect us with diseases such as Zika—would decrease. However, actually capturing, altering, and introducing mosquito populations into the larger mosquito community might be more difficult than enthusiastic media and scientists lead one to believe.

Another possible application of CRISPR in the environmental context involves a broader introduction of CRISPR-edited members of invasive species "with a deleterious trait (e.g., distorted sex ratio, reduced fertility, chemical sensitivity)" (Webber, Raghu, and Edwards 2015, 10567). The goal is for the edited members of the species to reproduce with non-edited members in the wild, causing the non-edited members to disappear from the environment over time. Although the application seems promising in the removal of invasive species that threaten eco-systems—such as the Zebra Mussel and the Giant African Snail (ibid., 10565)—commentators have expressed some concerns, including about the unforeseen consequences of introducing such organisms into the environment, such as the potential side-effect of completely wiping out an entire species from the planet (ibid., 10566).

How Do We Collect Data About Unintended Consequences of the Technology?

The unpredictability and complications of biotechnology are not always acknowledged by scientists and technicians. To help warn of the risks of biotechnology, artists are creating works that, while technically "successful" in the outcome of the modifications, nonetheless present a disadvantage for the species. Laura Cinti claimed to have modified a cactus to grow human pubic hair where it would otherwise have produced needles (Cinti n.d.; Waters 2009, 167). The cactus's needles serve as both protection from animals and birds that would prey upon it for food and hydration, and as conduits for much-needed moisture for the cactus itself. By replacing the cactus's main source of protection and hydration with the pubic hair of humans, *The Cactus Project* illustrates the diminution of an organism, here a cactus, through genetic alteration.

Such works highlight the possibility of unintended consequences, which are also possible with gene editing through CRISPR. For example, when researchers modified a mouse gene to protect against cancer, the modification caused premature aging (Tyner et al. 2002). Even a CRISPR intervention that seems like a panacea—modifying Zika mosquitoes to die out—may have unexpected and negative ecological consequences. Certain fish survive by eating mosquito larvae. In the Arctic tundra, where food sources are sparse, migratory birds feast on mosquito swarms. One ecologist estimated that if mosquitoes were eliminated, there would be a 50 percent loss of birds in the tundra (Fang 2010).

One group of activists believes CRISPR "may pave new ways to fight the ivory black market." This could include "engineering biomarkers into tusks to track poaching, or . . . alter[ing] tusks in a way that make them valueless to the ivory trade" (The Long Now Foundation n.d.). However, altering the tusks of elephants could bring about a series of unforeseen consequences. Changing the color or composition of the tusks might actually backfire and make the tusks more popular to consumers and more valuable to poachers. Any changes to the tusks could also affect the elephant's protective and reproductive capacities. Rubber-like tusks, for example, would most likely not be popular among consumers but would be effectively useless for protection, increasing elephants' susceptibility to other predators.

A fundamental change to the tusks could also make female elephants no longer interested in altered males. This would lead to one of two results: female elephants would only mate with the elephants with the original ivory tusks, furthering that original line and not the genetically altered one or, in the case that the genetic changes were carried out *en masse*, female elephants might not mate very much or even at all with their genetically altered male counterparts, decreasing the elephant population.

Life science artists' works point out a need for a mechanism to evaluate risks before the widespread use of a genetic technology. This is certainly true of CRISPR, where species that have been altered by gene drives entering the environment can do damage in unexpected ways. The FDA and USDA exert little regulatory oversight for CRISPR-altered species and products. Thus, there is little done to counteract what J. Benjamin Hurlbut has identified in this volume as the tendency to exaggerate benefits and minimize risks. As he points out:

> Advances in contemporary bioscience and biotechnology are often marked (and celebrated) as revolutionary: novel powers over life that portend profound changes to human lives and societies. At the same time, associated risks are rarely described in similar terms.
>
> (Chapter 4, page 77)

Who Owns Genetic Materials, Genetic Processes, and Genetic Information?

Life science artists and novelists raise concerns about the commodification of biological materials and the problems of intellectual property protections of biological processes and biological products. Artist Chrissy Conant exhibited Chrissy Caviar, tins of her ovum, in a New York gallery as a commentary upon ads soliciting egg donors for infertile couples and offering 50,000 to 100,000 dollars to those donors (Conant n.d.; Andrews 2007a, 136). Critical Art Ensemble's *Flesh Machine* was a 30-minute performance in which audience members took gamete donor screening tests. Those who passed the screening tests were asked to give blood for DNA extraction and amplification, which was carried out at an on-site lab. A profile of the participants was constructed, and the participants were told their potential value in the genetic market economy (Critical Art Ensemble n.d.).

When the California Supreme Court in *Moore v. Regents of the University of California* (1990) held that John Moore had no property right to his own cell line (which had been secretly patented by his doctor), artist Larry Miller expressed his dismay. Miller, struck by the questions of control and ownership of the body raised by the case, created a Genetic Code Copyright, an elegantly drawn certificate stating: "I . . . born a natural born human being . . . do hereby forever copyright my unique genetic code, however it may be scientifically determined, described or otherwise expressed" (Miller n.d.; Gessert 2010, 115). He thus challenged the idea that a person can be treated as an object—copyrighted, commodified, and patented.

Novelists, too, have raised concerns about the commodification of the biological processes and body parts through patent law. Michael Crichton's *Next* (2006) was inspired, in part, by the John Moore case. In the analogous court case in the novel, *Frank M. Burnet v. Regents of the University of California*, Frank sues for the unauthorized patenting of his cell line, but the court rules against him, finding that his cells were "waste" and not his property. The court even rules that the university can take cells from Frank and his descendants under the law of eminent domain. Ultimately, Frank's daughter Alex wins back the right to control the family's cell line by arguing that the company's ownership of cell lines will slow down research and violate the 13th Amendment, dealing with the abolition of slavery.

Similarly, in Richard Powers' *Generosity: An Enhancement* (2009), a biotech company tries to commercialize a woman's "happiness gene," to the point that it claims that her donation of her own eggs will violate its patent. In the fictional legal case in the book, *Truecyte v. Future Families Fertility Center, Houston*, the main character wins the right to control her eggs in their natural state, while the company wins the right to "a reasonable licensing fee for any novel tests or products" (ibid., 259).

Although the artists' and novelists' works predated the U.S. Supreme Court decision in *Association for Molecular Pathology v. Myriad Genetics, Inc.* (2013), their analyses predicted the logic that would be used in the actual U.S. Supreme Court decision. Although tens of thousands of genes had been patented by the time the case reached the Court, the Justices unanimously held that patents on human genes should never have been granted.

The basis for the United States patent system is the Progress Clause of the U.S. Constitution, which states that Congress shall have the power "to promote the Progress of Science and useful Arts, by securing for limited Times to Authors and Inventors the exclusive Right to their respective Writings and Discoveries" (U.S. Const. art. I, § 8, cl. 8). The Patent Act rewards inventors with a 20-year monopoly that forbids anyone else from making, using, or selling their invention in order to make sure that novel, non-obvious, and useful technologies get developed that otherwise might not have been created. Nature's handiwork is excluded from patentability (*Mayo Collaborative Services v. Prometheus Labs.* 2012; *Bilski v. Kappos* 2010). A newly discovered natural phenomenon must be "treated as though it were a familiar part of the prior art" and free for all to use (*Parker v. Flook* 1978; *Bilski v. Kappos* 2010). For 150 years, the Supreme Court had held that products of nature and processes of nature were not patentable—yet that precedent had been ignored by biotech companies and the patent office in the patenting of genes.

In striking down patents on human genes, the U.S. Supreme Court in *Association for Molecular Pathology v. Myriad Genetics, Inc.* (2013) wrote:

> Laws of nature, natural phenomena, and abstract ideas are not patentable. Rather, they are the basic tools of scientific and technological work that lie beyond the domain of patent protection. As the Court has explained, without this exception, there would be considerable danger that the grant of patents would tie up the use of such tools and thereby inhibit future innovation

premised upon them. This would be at odds with the very point of patents, which exist to promote creation.

With CRISPR, patents are front and center once again. Almost as much journalistic ink has been spilled on the battle between Jennifer Doudna's team and Feng Zhang's team for patent superiority as has been used to describe the scientific process itself. But, as with the patent on the gene sequence in *Myriad*, both patents parties' set of claims seem to be vulnerable to attacks based on the grounds of unpatentable subject matter. The Supreme Court has long held that natural products and processes are not patentable. CRISPR is based on a natural process by which RNA uses a protein to snip out a piece of DNA in the genome and then stitch the ends together (Sander and Joung 2004). A scientific process based on something that occurs in nature cannot be patented unless it is "markedly different" from the underlying process in nature (*Diamond v. Chakrabarty* 1980). The reason it is important not to have patents on products of nature or laws of nature, as Justice Breyer has noted, is that it would give inventors "*too much* patent protection" and "impede rather than 'promote . . . ' the constitutional objective of patent and copyright protection" (*Lab. Corp. of America Holdings v. Metabolite Labs., Inc.* 2006, 127).

In *Myriad*, the challenged patents claimed isolated BRCA1 and BRCA2 breast cancer genes, identified in the patent by their sequences—the long string of the chemical bases, A, C, G and T. The patents also covered any fragment of the BRCA1 gene sequence that was at least 15 nucleotides in length. Since those 15-nucleotide sequences occur at an average of 14 times per gene, Myriad could potentially have asked for a royalty on every test done on any gene. Myriad could thus hold hostage the deployment of whole genome sequence testing by threatening to pursue an infringement action for every instance one of those 15-nucleotide segments is sequenced. The patent claims on both the gene sequences and the 15-nucelotide segments of it were found to be invalid in the *Myriad* case. It did not matter that they were created in the lab because they were not markedly different from their naturally occurring counterparts.

The claims made in the CRISPR patent filings are equally problematic under patent law. The Doudna patent application initially claimed all 8-nucleotide segments of a larger RNA sequence, even though such segments exist in nature throughout the genome (see, for example, claims 2 and 45 of U.S. Patent Application No. 2014/0068797 2013). That patent application's claim 72 was even more overreaching—not just claiming an 8-nucleotide sequence, but all other nucleotide sequences that have at least 60 percent identity with each 8-nucleotide sequence of a longer sequence claimed by the patent.

As to the underlying CRISPR process, it would appear to be unpatentable under Supreme Court precedents. In *Funk Bros. Seed Co. v. Kalo Inoculant Co.*, the patent applicant isolated certain naturally occurring bacteria and combined them in a novel and useful way, yet this did not convert the bacteria from ineligible "phenomena of nature" to eligible inventions (1948). To permit the patent would have required "allowing a patent to issue on one of the ancient secrets of nature now disclosed." The CRISPR process, like each bacterium in *Funk Brothers*, "has the same effect it always had . . . [and] perform[s] in [its] natural way" (ibid., 131). "They serve the ends nature originally provided and act quite independently of any effort of the patentee" (ibid.).

The Zhang patent tries through a sleight-of-hand to gain patentability. That patent states in claim 1, with circular logic, that it covers "non-naturally occurring" CRISPR with certain characteristics (U.S. Patent No. 8,697,359 2014). In essence, the Zhang patent is trying to usurp the role of the legal system declaring what it covers to be non-naturally occurring and thus valid. But why should we believe a self-serving legal conclusion masked as a patent claim? Would we believe every criminal defendant who declared himself innocent? The claim seems more like a piece of performance art than a description of a patentable invention. However, unlike Stuart Newman's performance art chimera patent in Chapter 7, Zhang's patent was granted, conclusory language notwithstanding.

Although Doudna filed a patent application before Zhang did, Zhang paid an extra fee for an expedited review by the patent office and his patent was granted first. Doudna then filed an interference proceeding at the U.S. Patent and Trademark Office, claiming that Zhang's patent claims were not sufficiently different from hers for his patent to considered valid. In February 2017, the Patent and Trademark Appeals Board held that his patent claims did not overlap with hers (*Broad Institute v. Regents of the University of California* 2017).

As important as what was decided in their case is what was not decided. The decision did not analyze whether Doudna's patent application would be granted. She made immediate claims that she would be granted a broader patent than Zhang's, saying "They have a patent on green tennis balls; we will have a patent on all tennis balls" (Pollack 2017).

The decision also did not address whether the "inventions" claimed by either Zhang or Doudna were unpatentable biological processes. Such a claim can be brought in the future by other scientists—for example, those who wish to use forms of CRISPR without paying a royalty.

Allowing patents on the underlying CRISPR process is inconsistent with policies that have been part of the Patent Act since its inception. The Patent Act of 1793, drafted by Thomas Jefferson, stated that "simply changing the form or the proportions of any machine, or composition of matter, in any degree shall not be deemed a discovery."

How Can We Develop Genetic Technology with an Ethics of Caring and Responsibility?

Life science artists underscore that our genetic technologies and biotechnologies bring with them responsibilities. In using their artwork to contemplate the relationship between themselves and the living organisms they use as their medium, life science artists have introduced the concept "ethics of caring." Scientists are accustomed to relatively unfettered use of laboratory animals for years as they are most often focused on furthering scientific developments. Consequently, the ethics involved in the application of genetics and biological modifications to organic life does not cause a great stir in the laboratory community (see, e.g., discussion about the ethics of mosquito lab work; Chapter 3, this volume). Life science artists, however, typically focus simultaneously on both critiquing scientific methodologies, as well as reflecting on their own responsibilities to their living artworks.

Life science art underscores the role in the life sciences of caring—as well as killing. By creating art such as tissue culture art that is living and hence requires nurturing, Oron Catts and Ionat Zurr say they create an "aesthetics of caring" (Bureaud 2002). In the *Semi-Living Worry Dolls* exhibit, they built a bioreactor in the gallery and fed the cultured cell line throughout the exhibit, providing a greater understanding about what it would mean to bring a biotech product into the world. At the end of the exhibition, when they needed to shut down the bioreactor that was keeping their creations alive in the gallery, they engaged the audience in the decision of how to respectfully terminate life that has been brought about in the name of science.

Kathy High's *Embracing Animal* was a gallery exhibition that involved building a home for, and feeding, three retired lab rats (named Tara, Star, and Matilda Barbie) that had been genetically modified to be used for research on Crohn's disease, from which High herself suffers. High, a prospective beneficiary of the research, took on the subsequent care of the rats, feeling it was important to reciprocate the services received by their servitude. When she adopted the rats, they were bald and in pain. After her care and some homeopathic remedies, their fur grew back and they became more playful (Stracey 2009, 499; Thompson 2005, 60–67). Similarly, Eduardo Kac established guidelines for the care of his GFP bunny during conception. For Kac, it was important that the bunny not lose her health, her sociability with humans, or her interaction with other animals as a result of her genetic modifications. This led to the use of a specific type of GFP gene that ostensibly allowed otherwise "normal" development of Alba, the GFP bunny.

George Gessert, whose art includes breeding plants, suggests along these lines that:

> the intensity of the medium breaks the spell cast by traditional art, in which life seems to exist freed from death. No serious breeding project can indulge this illusion, because evolution, even on the aesthetic level, cannot occur without death.
>
> (2002, 41)

In *Art Life*, Gessert exhibited flowers of many different colors created by selective breeding. He asked the viewers to vote on their favorites. The rejected ones were turned into compost in front of the audience's eyes. He thus made visible to the public the type of "culling" that takes place within science—the number of plants, animals, products (and, in some instances, human embryos) that are rejected before an "acceptable" or "preferred" biological product is made available to the public. Gessert's approach contrasts starkly with Steichen's delphinium exhibit in the 1930s. As art historian Frances Stracey notes, Steichen omitted from the show "ugly, stunted, febrile rejects" (2009, 496). Gessert instead makes visible science's mistakes.

Sculpting Regulation

Life science artists provide a valuable perspective to help sculpt public policies regulating genetic technology. Artists have questioned the sources of information regarding the safety and application of genetic technology. They have challenged the current regulations, or lack thereof, that allow (or prohibit) the use of genetic information and genetic technologies. Some artists have imagined new uses of the technologies, forcing policy makers to think more creatively and with more foresight.

The issues raised by artists focusing on genetic and biological technology can challenge and catalyze the ongoing discussions taking place among the scientific and legislative communities. By mining their statements, their works, and the public response to the art, policy makers have the opportunity to receive valuable insight into the type of policies that may be appropriate for our genetic future.

Acknowledgments

I am grateful for the contributions of Michael Goodyear and Alexandra Franco to this chapter. I am also grateful for the personal conversations I have had with many life science artists, including Eduardo Kac, Oron Catts, Ionat Zurr, Kathy High, Orkan Telhan, George Gessert, Natalie Jeremijenko, Chrissy Conant, Larry Miller, and Adam Zaretsky.

References

Andrews, Lori. 2007a. "Art as a Public Policy Medium." In *Signs of Life: Bio Art and Beyond*. Edited by Eduardo Kac, 125–149. Cambridge, MA: The MIT Press.
———. 2007b. "Tissue Culture." *The Journal of Life Sciences* September: 68–73.
Association for Molecular Pathology v. Myriad Genetics, Inc. 580 U.S. __ (2013).
Bilski v. Kappos. 561 U.S. 593 (2010).
Broad Institute v. Regents of the University of California. 2017 WL 657415 (Patent Tr. & App. Bd. 2017).
Bureaud, Annick. 2002. "The Ethics and Aesthetics of Biological Art." *Art Press* 276: 39.
Cartmill, Cleve. 1944. "Deadline." *Astounding* 33: 154–178.
Chin, Mel. n.d. "Revival Field." Available at: http://melchin.org/oeuvre/revival-field
Cinti, Laura. n.d. "The Cactus Project." Available at: http://thecactusproject.com
Comis, Don. 1995. "Metal-Scavenging Plants to Cleanse the Soil." *Agricultural Research* 43 (11): 4–9.

Conant, Chrissy. n.d. "Chrissy Caviar." Available at: www.chrissycaviar.com/ccaviar

Crichton, Michael. 2006. *Next*. New York: HarperCollins.

Critical Art Ensemble. n.d. "Flesh Machine." Available at: http://critical-art.net/?p=101

Cyranoski, David. 2015. "Gene-Edited 'Micropigs' to Be Sold as Pets at Chinese Institute." *Nature* 526: 18.

———. 2016. "CRISPR Gene Editing Tested in a Person." *Nature* 539: 479.

Diamond v. Chakrabarty. 447 U.S. 303 (1980).

Essaïdi, Jalila et. al. 2012. *Bulletproof Skin: Exploring Boundaries by Piercing Barriers*. Jalila Essaïdi.

Fang, Janet. 2010. "A World Without Mosquitos." *Nature* 466: 432–434.

Friko, Michael J. 2016. "Re: Request for Confirmation That Transgene-Free, CRISPR-Edited Mushroom is Not a Regulated Article." *United States Department of Agriculture*. April 13. Available at: https://www.aphis.usda.gov/biotechnology/downloads/reg_loi/15-321-01_air_response_signed.pdf

Funk Bros. Seed Co. v. Kalo Inoculant Co. 333 U.S. 127 (1948).

Gambino, Megan. 2013. "The Story of How an Artist Created a Genetic Hybrid of Himself and a Petunia." *Smithsonian Magazine*. February 22. Available at: www.smithsonianmag.com/science-nature/the-story-of-how-an-artist-created-a-genetic-hybrid-of-himself-and-a-petunia-25148544

Gedrim, Ronald J. 1993. "Edward Steichen's 1936 Exhibition of Delphinium Blooms: An Art of Flower Breeding." *History of Photography* 17: 352–363.

Gessert, George. 2002. "Plants and the Art of Evolution." *Art Press* 276: 41.

———. 2010. *Green Light: Toward an Art of Evolution*. Cambridge, MA: The MIT Press.

Gordon, Jon W. 1994. "Genetic Enhancement in Humans." *Science* 283: 2023–2024.

Gordon, Linda. 2009. *Dorothea Lange: A Life Beyond Limits*. New York: W.W. Norton & Company.

Jacobs, Thomas B. et al. 2015. "Targeted Gene Modifications in Soybean with CRISPR/Cas9." *BMC Biotechnology* 15: 1–10.

Jeremijenko, Natalie. 2007. "OneTree." In *Signs of Life: Bio Art and Beyond*. Edited by Eduardo Kac, 301–302. Cambridge, MA: The MIT Press.

Jia, Hongge and Nian Wang. 2014. "Targeted Genome Editing of Sweet Orange Using Cas9/sgRNA." *PLOS ONE* 9: e93806.

Jones, Steve. 2015. "How Not to Clone a Mammoth." *The Lancet* 386: 125.

Kac, Eduardo. n.d. *"Genesis: A Detailed Description of Genesis."* Available at: www.ekac.org/geninfo2.html

———. 1998. "Transgenic Art." *Leonardo Electronic Almanac* 6. December. Available at: http://mitpress.mit.edu/e-journals/LEA

———. 2005. *Telepresence & Bio Art: Networking Humans, Rabbits, & Robots*. Ann Arbor, MI: University of Michigan Press.

Krisch, Joshua A. 2016. "This Gene-Editing Tool Could Destroy Zika Virus." *Vocativ*. February 17. Available at: www.vocativ.com/286508/crispr-zika

Lab. Corp. of America Holdings v. Metabolite Labs., Inc. 548 U.S. 124, 127 (2006) (J. Breyer, dissenting).

The Long Now Foundation. n.d. "Revive & Restore; Genetic Rescue for Endangered and Extinct Species; Woolly Mammoth Revival." Available at: http://reviverestore.org/projects/woolly-mammoth

Mayo Collaborative Services v. Prometheus Labs. 566 U.S. __ (2012).

Meltzer, Milton. 2000. *Dorothea Lange: A Photographer's Life*. Syracuse, NY: Syracuse University Press.

Mikami, Masafumi, Seiichi Toki and Masaki Endo. 2015. "Comparison of CRISPR/Cas9 Expression Constructs for Efficient Targeted Mutagenesis in Rice." *Plant Molecular Biology* 88: 561–572.

Miller, Larry. n.d. "Larry Miller." Available at: www.onlyonelarrymiller.com

Moore v. Regents of the University of California. 793 P.2d 479 (Cal. 1990).

National Academies of Sciences, Engineering, and Medicine. 2016. *Gene Drives on the Horizon: Advancing Science, Navigating Uncertainty, and Aligning Research with Public Values.* Washington, D.C.: The National Academies Press.

Parker v. Flook. 437 U.S. 584 (1978).

Patent Act of 1793. Ch. 11, § 2, 1 Stat. 318–23 (February 21).

Pollack, Andrew. 2017. "Harvard and M.I.T. Scientists Win Gene-Editing Patent Fight." *The New York Times.* February 15. Available at: https://www.nytimes.com/2017/02/15/science/broad-institute-harvard-mit-gene-editing-patent.html?_r=0

Powers, Richard. 2009. *Generosity: An Enhancement.* New York: Farrar, Straus, and Giroux.

Reardon, Sara. 2016. "First CRISPR Human Clinical Trial Gets a Green Light from the U.S." *Scientific American.* June 22. Available at: www.scientificamerican.com/article/first-crispr-human-clinical-trial-gets-a-green-light-from-the-u-s

Rodriguez, E. 2016. "Ethical Issues in Genome Editing Using CRISPR/Cas9 System." *Journal of Clinical Research & Bioethics* 7: 266.

Royal Commission on New Reproductive Technologies. 1993. *Proceed with Care*, vols I and II. Ottawa, ON: Minister of Government Services.

Sander, Jeffry D. and J. Keith Joung. 2004. "CRISPR-Cas Systems for Editing, Regulating and Targeting Genomes." *Nature Biotechnology* 32: 347–355.

Sinclair, Upton. 1906. *The Jungle.* New York: Knopf Doubleday Publishing Group.

Stein, Laura. n.d. "Past Works." Available at: www.laurastein.com/home

Stracey, Frances. 2009. "Bio-art: The Ethics Behind the Aesthetics." *Nature Reviews Molecular Cell Biology* 10: 496–500.

Strickland, Carol. 2014. "Getting the Lead Out: Mel Chin." *Art in America.* April 1. Available at: www.artinamericamagazine.com/news-features/interviews/getting-the-lead-out-mel-chin

Tang, Ya-Ping et al. 1999. "Genetic Enhancement of Learning and Memory in Mice." *Nature* 401: 63–69.

Thompson, Nato, ed. 2005. *Becoming Animal: Contemporary Art in the Animal Kingdom.* North Adams, MA: MASS MoCA Publications.

Tyner, Stuart D. et al. 2002. "P53 Mutant Mice that Display Early Ageing-Associated Phenotypes." *Nature* 451: 45–53.

U.S. Const. art. I, § 8, cl. 8.

U.S. Patent Application No. 2014/0068797. 2013. "Methods and Compositions for RNA-Directed Target DNA Modification and for RNA-Directed Modulation of Transcription." Filed March 15, 2013.

U.S. Patent No. 8,697,359. 2014. "CRISPR-Cas Systems and Methods for Altering Expression of Gene Products." Filed October 15, 2013. Issued April 15, 2014.

Waltz, Emily. 2016. "Gene-Edited CRISPR Mushroom Escapes US Regulation." *Nature News* 532: 293.

Wang, Shaohui et al. 2015. "Efficient Targeted Mutagenesis in Potato by the CRISPR/Cas9 System." *Plant Cell Reports* 34: 1473–1476.

Wang, Yangpeng et al. 2014. "Simultaneous Editing of Three Homoeoalleles in Hexaploid Bread Wheat Confers Heritable Resistance to Powdery Mildew." *Nature Biotechnology* 32: 947–951.

Waters, Brent. 2009. *This Mortal Flesh: Incarnation and Bioethics*. Grand Rapids, MI: Brazos Press.

Webber, Bruce L., S. Raghu and Owain R. Edwards. 2015. "Opinion: Is CRISPR-Based Gene Drive a Biocontrol Silver Bullet or Global Conservation Threat." *Proceedings of the National Academy of Sciences* 112: 10565–10567.

Wei, Feng et al. 2001. "Genetic Enhancement of Inflammatory Pain by Forebrain NR2B Overexpression," *Nature Neuroscience* 4: 164–169.

Wunderman, Ali. 2016. "Our Love Affair with Purebred Dogs Has Created Genetically Inferior Animals." *Quartz*. January 11, http://qz.com/590678/our-love-affair-with-purebred-dogs-has-created-genetically-inferior-animals/

Youngs, Amy. n.d. "Techno-Eco-Engineering." Available at: www.ylem.org/artists/ayoungs/techno-eco.html

Zhou, Xiaohong et al. 2015. "Exploiting SNPs for Biallelic CRISPR Mutations in the Outcrossing Woody Perennial *Populus* Reveals 4-Coumarate: CoA Ligase Specificity and Redundancy." *New Phytologist* 208: 298–301.

Part III

Human-Nonhuman Boundaries, Worked and Reworked

Chapter 7

Sex, Lies, and Genetic Engineering

Why We Must (But Won't) Ban Human Embryo Modification

Stuart A. Newman

Introduction

As any technology becomes more powerful and versatile there is an inclination to accept claims that it can be used to make things better. In the case of biotechnologies for modifying organisms, leaps in capability and precision are frequent, and the impulse to use them to improve human health and socially desirable features is culturally ingrained. In biology, however, things are rarely straightforward. For example, while cloning, a means of producing animals with identical genomes, is now used in the cattle industry, the decades-old prospect of using it to produce better people (Myhrvold 1997; Staff 1997) has dwindled, as it has become clear that vastly more than genes needs to be controlled to achieve desired biological outcomes (Leonard 2016).

The sequencing of the human genome coupled with low-cost means of diagnosing potentially deleterious gene variants has been touted as a way to eliminate gene-related disease and eugenically advantage one's children. A recent book by a Stanford University bioethicist, for example, eagerly anticipates a world where the making of humans will no longer be mediated by sex, but by selection among genetically characterized *in vitro* embryos, perhaps touched up using the newly developed CRISPR-Cas9 technology (Greely 2016). A Harvard University genetic technologist with a foot in both academia and industry and a strong presence on the public stage, has been less reticent about promoting the imagined benefits of a genetically engineered human future (Church and Regis 2012).

Whether or not it can be accurately performed, the disassembly, reassembly, and reconfiguration of human organisms is already underway. The relevant methods are being developed with anticipation of great wealth from the sale of new products. These could be replacement tissues and organs, model systems for testing the effects of new drugs or, more ambitiously, children with biological features that their parents are otherwise unable to provide. As with other organisms and their tissues, human parts and part-humans are now raw materials of the industrial system. And similarly to other high-stakes manufacturing and commercial ventures, there has been little democratic control over these technologies and no shortage of academic and other opinion makers (many of them financial beneficiaries) set on normalizing them and diverting attention from their potentially radical civilizational implications.

Modification of an existing person's biology to save a life or alleviate suffering, no matter how new or unprecedented the methods, is clearly in the traditional province of medicine. In contrast, modifying the biology of someone who has not yet come into existence, to prevent malfunctions or enhance desired qualities, is outside medicine's classic mission. While there are many reasons why prenatal manipulation of humans is hazardous and unethical, such considerations have become largely irrelevant to preventing it. There was a window of time after World War II when the reality of a eugenicist social agenda was still a raw memory (but the techniques for prenatal manipulation had not yet come into being) when preemptive alteration of human biology might have been proscribed. That opportunity, if it was ever a real one, has passed.

It is not that modern societies lack taboos. Nearly universal prohibitions against incest and cannibalism are consensually enforced throughout much of the world, and are the stuff of scandal when breached. Opponents of some other activities—slavery, abortion, capital punishment, torture—seek to place them in this category as well. Modifying embryos with the intention of bringing them to term would seem to be an ideal cause for prohibition on moral grounds, considering that it is a hazard to subjects who have no say over what is done to them.

The Declaration of Helsinki (WMA 2013), first promulgated in 1975, contains international agreements on research involving human subjects. It recognizes that human research of potential benefit to the subject (e.g., on volunteers) is preferable to that which is not, and makes special provisions for subjects who are incompetent and who are incapable of giving consent. It significantly relaxed the provisions of the post-World War II Nuremberg Code (U.S. Government 1949), which held that informed consent was "absolutely essential." From that point, only the consent of a surrogate acting in the subject's interest was required. But while a sick individual has obvious interests, a prospective human (a category of subject not anticipated by either the Nuremburg or Helsinki codes) does not. Anyone who seeks to reconfigure the developmental fate of a human embryo has their own reasons to do so. These reasons, by definition, cannot be judged against the needs of someone who does not yet exist (Habermas 2003).

Although many prominent scientists and bioethicists are sanguine about the prospect of genetic modification of future children (Church and Regis 2012; Baltimore et al. 2015), there is resistance to the idea from broad sectors of the public (STAT-Harvard 2016) and from a variety of scientific, ethical, and social equity perspectives (Lippman et al. 1993; Billings, Hubbard, and Newman 1999; Annas, Andrews, and Isasi 2002; Darnovsky 2008; Lanphier et al. 2015; Subbaraman 2016). As I will show, it would be infeasible to prevent genetic modification of human offspring without stopping all modification of human embryos, even where there is no intention of bringing them to term, but very few secular individuals or organizations have called for such a ban (Alliance for Humane Biotechnology n.d.). One reason for this is that scientific and medical researchers see the embryo as a material ripe for investigation (Peura, Schaft, and Stojanov 2010; Bahadur et al. 2008). While most biologists recognize the profound uncertainties that contraindicate genetic manipulation and cloning, they can obtain benefit even from experiments that turn out unexpectedly or badly.

Those supporting embryo research only for scientific and medical purposes believe that the regulatory process can be used to ensure that experimental embryos will not be permitted to develop to term (Brokowski, Pollack, and Pollack 2015; Comfort 2015). But in countries where there are statutory limits on how long experimentally manipulated human embryos are permitted to survive (for instance, a 14-day cutoff has been in place in the U.K. since 1990; HFEA 2016), such limits can be extended when health benefits (even speculative ones) enabled by new technological advances are brought to bear (Hyun, Wilkerson, and Johnston 2016). Public perception of the relevant scientific issues is under heavy influence from corporations (Philpott 2012; Ruskin 2015) whose interests are represented by surrogates such as Americans for Cures (n.d.) and the Genetic Literacy Project (n.d.). The information disbursed by these organizations is often inaccurate (Stevens 2007).

Apart from the promise of fame and fortune, there are ideological factors that potentially compromise the policy advice of experts. Though called into question or disconfirmed in recent years, gene-centric explanations (as in genes "for" obesity, diabetes, depression) (Krimsky and Gruber 2013) tend to be reflexively invoked when justifying new methodologies and treatments. While the increasing precision of making genetic changes by CRISPR is frequently used as a selling point for the prospect of "correcting" human embryos, with the recognition that genes do not work in a consistent way across all animals (True and Haag 2001), or even between individuals of the same species (Narasimhan et al. 2016), the relevance of such precision is significantly lessened.

The field of bioethics, particularly in the English-speaking world, has been accommodating to the received genetic determinism. The members of this discipline who write about reproductive biotechnologies ("reproductive bioethicists"), apart than those writing from a Catholic perspective (Byrnes 2005; George 2016; Hurlbut 2005a; Pacholczyk and Hurlbut 2005) hold an array of positions ranging from laissez-faire libertarianism to technophilic liberalism. As documented in detail by the historian M.L. Tina Stevens (Stevens 2000), the field includes very few who problematize hegemonic societal agendas and power relationships or have a critical perspective on science as a social activity. In response to both perceived and genuine lack of scientific understanding by the public, reproductive bioethicists, with some rare and notable exceptions, see their responsibility as the quelling of anxiety about new developments by explaining them in the engineering language favored by many of their scientific colleagues.

In some cases, bioethicists have issued warnings about the negative potential of such alterations, such as the exacerbation of inequality or the devaluation of the congenitally impaired (Singer 2003; Evans and Moreno 2015), though without challenging the capability of scientists to make the individual improvements they propose. However, even this is too much for some commentators. The psychologist Steven Pinker considers these mild caveats and the nod to democratic process as inimical to science and medicine, and has called for bioethicists to "get out of the way" (2015).

The lack of skepticism by reproductive bioethicists sustains their value as solid academic citizens—unthreatening to their colleagues' research or business ventures, mindful of their institutions' bottom line. It has garnered remunerated seats on boards of biotechnology companies for some, as well as appointments to federal advisory panels (Elliott 2001). While it is understandable that these scholars would generally subscribe to the dominant biological ideologies in these fast-moving fields, in some cases (as in the ongoing controversy about three-parent embryos constructed by mitochondria-directed egg modification; see below), they go well beyond deferential reticence to knowingly giving cover to inaccurate and misleading narratives. Science journalists and popularizers rarely do any better than the bioethicists in critiquing the assumptions of their scientist sources, whose access they would likely lose if they pressed too hard. There are some notable exceptions (Dowie 2004; Stein 2016; Begley 2016).

In the next sections of this chapter, I will describe how policy recommendations regarding proposed human biological modifications have been arrived at by sidestepping accepted scientific knowledge that is well documented in popular textbooks or in easily accessible research reports. These cases involve clear derelictions of responsibility by bioethicists that can be (only) partly laid on the scientists who serve as their sources. I will conclude by speculating on the consequences of the prospect (which I consider likely) that these biological issues will be ignored.

Confusions around Cloning

In 1997, Ian Wilmut and his Roslin Institute colleagues in Scotland announced the birth of a cloned mammal. To create the sheep they called Dolly, the nucleus of a fresh ovum (egg) taken from one adult sheep was removed and into it was inserted the nucleus of a mammary cell taken from another adult sheep (Wilmut et al. 1997). The cloning of a mammal was based on earlier research on frogs by Robert Briggs and Thomas J. King in the 1950s with embryo nuclei (Briggs and King 1952), and by John Gurdon in the 1960s with tadpole and adult nuclei (Gurdon 1968). Before the mid-twentieth century, no clones produced by somatic cell nuclear transfer (SCNT) had occurred in the history of life on Earth, at least not among vertebrate animals.

Since SCNT cloning does not benefit from the error-correcting mechanisms that have evolved over vast periods of time in response to the kinds of cellular damage normally experienced by organisms (such as errors in the replication of DNA), it is not surprising that the cloning of Dolly was preceded by hundreds of failed attempts. SCNT is a very low-efficiency procedure, accompanied by extensive fetal loss and congenital defects in newborns (Fulka and Fulka 2007). While a minority of clones can slip through the shakeout period and survive to old age, most do not, and the health prospects for clones vary with the species (Sinclair et al. 2016). Although SCNT is therefore unsuitable for making human beings, the fantasy of producing a new person from a genetic prototype caught the public's interest. As reports of Gurdon's work entered the culture in the 1970s, it was subject to various literary treatments, to comic effect, for example in Woody Allen's *Sleeper* (1973), and as a dystopian portent (cloning Hitler) in *The Boys from Brazil* (Levin 1976; *The Boys from Brazil* 1978).

There is, of course, another biological phenomenon that produces "clones," one that is much more familiar than, and very different from, SCNT. This is the dissociation of the cells of an early embryo (blastomeres) to produce "identical" (monozygotic) twins, triplets, quadruplets, and so forth. The upper limit to the number of genetically identical individuals produced naturally in this fashion seems to be 16, after which mammalian blastomeres are no longer *totipotent* (capable of producing all the embryonic and relevant extra-embryonic tissues). Even embryos that do not disaggregate in the normal course of gestation can be forced to do so. "Embryo splitting," which results in twinning of a desired mating outcome, has long been used in the cattle industry. In his 1931 novel *Brave New World* (1932), Aldous Huxley contemplated human applications in the "Bokanovsky Process," by which the cellular products of a single fertilized ovum were subdivided to form up to 96 identical embryos. These were treated in specific ways to produce humans of the lower Gamma, Delta, and Epsilon classes. The higher-order Alpha and Beta humans, in contrast, were permitted to develop without modification.

SCNT clones have little in common, biologically, with human twins or with the cohorts from Huxley's fictional "bokanovization" procedure. In the first place, unlike the genes in separated blastomeres of an embryo, which function in egg cytoplasms with identical composition, the genes of a SCNT clone operate in different egg environments and each additional SCNT clone of the same donor will develop in still different egg environments. While it is commonplace that monozygotic siblings (monozygotics) can be very different from one another in personality, abilities, health, and so forth, despite having the same egg environment during early development, SCNT-produced organisms are even more different from one another and from their genetic antecedent, the nucleus donor.

Second, natural identical twins do not have a genetic antecedent. Monozygotics (even *Brave New World*'s) may have identical genes, but they are all genetically unprecedented. Yet the impulse to create SCNT mammals (like the cloning of a deceased pet) is tied to a desire for the new animal to be as much as possible like a previous one. While those who undertake this procedure may recognize that genetic replicas are not exact copies, the activity entails an attempt to deny the new individual the biological uniqueness (a form of ontological autonomy) that pertained to each previous member of its species before the appearance of SCNT.

A third difference between monozygotics and SCNT clones—the most important one biologically—is that unlike the former, the latter are assembled from portions of severely damaged cells, i.e., an isolated nucleus and an egg that has had its own nucleus removed. It is a scientific mystery as to how these cell fragments can recover from the insult and occasionally cooperate to produce an apparently fit member of the respective species: this particular kind of biological train wreck has never occurred in the course of pre-biotech evolution.

The profound differences between SCNT clones and twins did not stop various influential commentators and scientific popularizers such as Richard Dawkins (1998) and Stephen Jay Gould (1997, 1998) from using the twin analogy to naturalize the prospect of SCNT cloning. To do so, they focused on supposed fears that clones would not be unique individuals. Gould noted that biological similarities between twins are even greater than between a SCNT clone and its progenitor (Gould 1998). Here he appealed to the different egg microenvironments and subsequent life experiences to which the shared SCNT genome is subjected (see difference 1, above). Gould thus wrote as if what was bothering critics about the prospect of producing SCNT humans was the fact that two humans would have the same genome (a straw man), rather than the bizarre motives of those who might wish to produce a person from a *preexisting* prototype (difference 2, a sociological issue), or the similarly crackpot enterprise of trying to make a human by putting pieces of cells together and hoping that nature will make it right (difference 3, a safety issue).

The two decades since Dolly's birth have seen enough health problems with cloned animals, as well as evidence that SCNT clones do not reliably express the desired qualities of their prototypes, to have taken the technique off the front burner of cutting-edge human reproductive methods (Leonard 2016). The bioethicist Gregory Pence of the University of Alabama has made a career of promoting SCNT using the twin analogy, producing a string of books with titles like *Who's Afraid of Human Cloning?* (1998), *Cloning After Dolly: Who's Still Afraid?* (2004), and *How to Build a Better Human* (2012), in which he referred to opponents of generating children by SCNT cloning as "reactionaries" and "Alarmists." His most recent book on this subject, *What We Talk About When We Talk About Clone Club* (2016), a disquisition on the Canadian television program *Orphan Black* (BBC America n.d.), represents a retreat from such name-calling in the face of scientific reality, apparently instilled through the program's compelling narrative. Here Pence writes (with no apologies):

> a high risk of creating babies with chromosomal defects, as well as other defects caused by genes not turning on or off at the right times during gestation . . . is the major reason why originating babies by cloning is unethical. There is a very high likelihood that any babies so produced would have major structural abnormalities, problems caused by deep-down irregularities in their genes and cells.

(Pence 2016)

Finally, it is obvious that twins are humans. But are SCNT clones necessarily so? According to the tenets of genetic determinism, if they have a species' genome, they are members of that species. Even if we concede this, what is the status of an animal developing from a sheep's egg containing a human nucleus?[1] Other plausible, scientifically motivated SCNT manipulations might include swapping out one or more of the 23 different human chromosomes for one from a nonhuman animal (Pereira et al. 2008), or substituting segments of human DNA with synthetic DNA sequences (Liskovykh et al. 2015). The way a SCNT clone is constructed enables such possibilities. Moreover, other than restrictions on federal funding, no laws in the United States restrict such efforts, even to the extent of bringing such engineered clones to term.[2]

Three-Parent Embryos

Techniques for generating infants who would inherit only nuclear genes from women who carry impairing mutations in their mitochondria (the cell organelles that extract energy from fuel molecules) were approved for clinical use by the British Parliament in early 2015 (Vogel 2015). They also were given the go-ahead by the FDA a year later (Reardon 2016). Mutations in mitochondria genes can adversely affect hearing, vision, pancreatic function, and neuromuscular activity, among other physiological systems.

The two approved methods involve inserting a cell nucleus isolated from the egg of one woman into an enucleated (nucleus-lacking) egg of another woman, either before ("maternal spindle transfer" or MST) or after ("pronuclear transfer" or PNT) fertilization (Wolf, Mitalipov, and Mitalipov 2015). All throughout the period of deliberation around their approval, these techniques were referred to as mitochondrial "transfer" or "replacement" by their scientist-creators, journalists, bioethicists, and members of regulatory panels, by legislators (Wolf, Mitalipov, and Mitalipov 2015; Chinnery et al. 2014; Ishii 2014; Ridley 2015), and even by some critics of the procedure (Dickenson 2013; Baylis 2013). Because no transfer of mitochondria is involved in the procedures, these descriptions are scientifically inaccurate. The bioethicist Nita Farahany, a member of President Obama's Bioethics Commission, curiously described it in the *Washington Post* as exactly the opposite of what it is: "Put simply, the mitochondria of the affected egg are removed and replaced by the mitochondria from the healthy egg" (2014).

In fact, both MST and PNT are similar to cloning by SCNT, described above. Like cloning, the techniques involve replacement of an egg's nucleus by a nucleus from another cell. In cloning, the transferred nucleus is from a cell of a fully developed animal (potentially, a human), making the resulting organism a genetic "copy" of the nucleus donor. In MST and PNT, the genotype of the transferred nucleus is novel, so that the resulting child will be genetically unprecedented.

Since it is true that nuclear genes of an affected woman will eventually coexist in the prospective child with mitochondria from a second woman, from the affected woman's viewpoint the mitochondria of the egg are different from her own—"replaced" in the same way one's windows are replaced when he or she moves into a new house. In addition to mitochondria, the second woman also provides all the non-nuclear components of her egg, which are well known to play major roles in determining the characteristics of the future child. The fact that the egg which will be implanted in someone's uterus (possibly of a woman different from the two cell-fragment donors) originated in the body of the second woman means that if *she* had legally contracted to be the baby's mother the manipulation would have been characterized as a "genome transfer" or "genome replacement." The most neutral characterization of the procedure is that an infant is produced using an egg that has had its nucleus (including its nuclear genes) removed and replaced by a nucleus obtained from a different person's egg.

Because a different woman from the egg producer provides an egg nucleus with its 20,000-plus genes, culturally entrenched genetic determinism makes it extremely easy for researcher-advocates of the three-parent procedures and their bioethicist- and popularizer-enablers to (i) represent the nucleus donor as the "mother," and (ii) misrepresent the nuclear transfer procedure as the replacement of a few genes and thus characterize the technique as a trivial manipulation. Thus, as reported in the journal *Nature*, the technique's developers "have compared mitochondrial replacement to changing the batteries in a camera" (Callaway 2014). Also, the writer Matt Ridley stated in his popular blog that the mitochondrial DNA "is 500 . . . times less in quantity than the DNA that runs the rest of the body and contains only 37 of our 22,000 genes. It is barely any more relevant to your personality than the bacteria in your gut" (but see Christian et al. 2015 about the relevance of bacteria to one's personality). Ridley continued: "To call somebody with donated mitochondria a three-parent child is like calling somebody with a kidney transplant a three-person chimera, for the gene ratio is the same: 0.2 per cent" (2015). In fact, all the nuclear genes and the vast majority of their variant versions (alleles) are shared by both the women (as well as all humans), so the nucleus donor's distinctive contribution to the offspring is many times less than Ridley claims.

The main importance of whether the procedure is represented as the replacement of a few dozen genes or tens of thousands is that the technology has been shown to be unsafe in animal studies, leading in its own right to congenital anomalies due to incompatibilities between nuclei and mitochondria that have not coevolved (Gershoni et al. 2014). This presented potential difficulties for public perception and the machinations of the regulatory process. Anyone who consults the testimony before the British Human Fertilisation and Embryology Authority (HFEA), the record of public response to these deliberations, and the debate before Parliament that eventually certified the three-parent procedure for clinical use, will be impressed by the effectiveness of the framing of the procedure ("37 genes"; "replacing batteries") in surmounting any opposition. Pitching it in terms of "20,000 genes" or "cloning" would not have played nearly as well.

More candid commentators do not deny scientific problems with these oversimplified, genetic determinist views, though they may have reasons of their own to minimize their effects. The bioethicist César Palacios-González acknowledges that the "term 'MRTs' [mitochondrial replacement therapies] is a misnomer, since we are not transferring or replacing mitochondria," but advises that anyone who considers that "using the acronym MRTs has a harmful propagandistic effect . . . mentally replaces it for the less alluring term 'Nuclear DNA Replacement Techniques Employed to Avoid mtDNA Disease'" (Palacios-Gonzalez 2016).

Such blandishments by proponents of the three-parent procedure in the run-up and subsequent to regulatory authorization in the U.K., and similar attempts in the United States, were proved successful when on September 26, 2016, a New York fertility physician reported accompanying a Jordanian family to Mexico, where he performed five instances of MST, leading to the birth of a baby boy (Hamzelou 2016). Apparently, no legal statutes in the United States, Jordan, or Mexico prohibited this arrangement. Four of the embryos failed to develop, or developed abnormally, raising questions about possible covert procedure-induced impairments in the ostensibly successful case.

Human–Nonhuman Chimeras

The previous two examples provide cautionary tales about how human species identity and biological integrity can be destabilized by advances in technology. Cloning, despite all the normalizing twin-talk that has surrounded it, is a deconstructive and reconstructive methodology. It can be put to boundary-crossing scientific and medical uses (e.g., human nuclei in sheep eggs, brain-deficient human clones for organ harvesting), which could readily erode the culturally significant, but biologically tenuous, human–nonhuman and human–humanoid distinctions. Generation of three-parent embryos, despite being another form of nuclear transfer that involves melding of severely damaged cells, has gained regulatory approval for the making of babies in the U.K. by being deceptively marketed ("mitochondrial replacement," "mitochondrial transfer") and portrayed as trivial and innocuous ("37 genes"). The speculative promise of preventing birth defects went unchallenged in most popular and academic discussions. Those who got wind of the game were ignored, scolded, or advised to look the other way (Pinker 2015; Palacios-Gonzalez 2016). Then, immediately after governmental approval of one of the techniques in the U.K., its developer started a campaign (and a company) to promote the use of the nuclear transfer technique for general reproductive purposes, not just disease prevention (Connor 2015).

With human-animal chimerism (recently approved for federal funding in the United States; see below), it has become impossible to avoid confronting the transgressive nature of slated technological interventions. Though the earliest work had been done with different species of mice, the technique was proved in a more dramatic fashion with the goat–sheep chimeras ("geeps") reported in 1984 (Meinecke-Tillmann and Meinecke 1984; Fehilly, Willadsen, and Tucker 1984). Embryo chimeras are different from hybrids (like mules) produced by cross-fertilization. The cells used to construct the chimeras retain their species identity, so that the resulting animals are intermediate in expressed biological features (phenotype) between the cell-donor species. A geep does not look exactly like a sheep or a goat, but looks somewhat like each. Since each egg or sperm of a female or male chimera has one or the other species identity, two geeps could mate and produce a sheep or a goat. Correspondingly, two pig–human chimeras could potentially birth a human. Interspecies embryo chimeras are very different from organisms that have received grafts of tissues of another species, such as human cardiac patients with pig valves. In these cases, engraftment takes place after early development of the host species is completed and its species character, including the organization of its nervous system and brain, has been set.

In 1997, the present writer applied for a patent on chimeric embryos and animals that would contain both human and nonhuman cells. The intention was not to create such organisms, but, by raising questions about the difficulties of drawing consistent and defensible lines between what is a human being and what is not, and what is a life form and what is an invention, to alert a broader public as to where these new technologies could lead (U.S. Patent Application No. 08/993,564 1997; see also Dowie 2004; Newman 2006). The application included claims for organisms containing combinations of human and mouse cells, as well as human and chimpanzee cells, with the human contribution to the outcome anywhere from a minor to a major proportion. The descriptions of utility included uses of partly human embryos and full-term animals in developmental biological research, as sources of transplantable tissues and organs for human patients, and for testing the toxicity of drugs and chemicals.

The ploy, though eliciting numerous law review articles dealing with the Constitutional issues it raised (e.g., Coughlin 2006; Heled 2014; Magnani 1999), caused significant consternation the legal and scientific community, even though the motivation was publicly acknowledged.[3] The chair of the department of genetics at Harvard Medical School, the scientist who patented the first mammal, told an interviewer: "[t]he creation of chimeras is an outlandish undertaking. No one is trying to do it at present, certainly not involving human beings" (Zwerdling 1998). In 1999, the U.S. Patent and Trademark Office (USPTO) rejected the invention on the grounds that some of its embodiments would "embrace a human being" (Weiss 1999), and held to this position through a series of responsive amendments until the final submission in 2005 (Weiss 2005; Newman 2006).

In the absence of a consensual taboo on such activities, however, the logic of science as currently practiced dictated that they would proceed. In 2006, a research group at Rockefeller University published a paper titled "Contribution of Human Embryonic Stem Cells to Mouse Blastocysts" (James, Noggle, Swigut, and Brivanlou 2006), and in 2007 a University of Nevada group created a sheep with partly human organs by injecting human stem cells into a sheep's fetus (Almeida-Porada et al. 2007). In 2009, the NIH published guidelines that prohibited funding for the breeding of chimeric animals in which human stem cells could become eggs or sperm (NIH 2009). Around the same time, the NIH also declined to fund experiments involving chimeras between humans and nonhuman primates, and it reiterated its position in 2015 (NIH 2015).

Funding restrictions are not bans, however, and some of the proscribed research nonetheless went forward with non-NIH support, including money from the state-funded California Institute of Regenerative Medicine (CIRM n.d.). CIRM was created in 2004 to support research with human embryo stem (ES) cells (Longaker, Baker, and Greely 2007; Trounson, Klein, and Murphy 2008). But after 2006, through the efforts of Shinya Yamanaka, induced pluripotent stem cells (iPSC) became the less controversial alternative (Yamanaka and Takahashi 2006). These cells, the functional equivalent of the blastomeres of the early embryo, are produced by activating a small number of developmentally early-acting genes in somatic (non-reproductive) cells of the mature body. While iPS cells cannot form human embryos on their own, they can generate any portion or percentage of a chimera (the human component) if inserted into an actual embryo, either human or animal.

Thus, while the *origin* of iPS cells is not in a human embryo (allowing them to escape the funding-threatening attention of "pro-life" activists and legislators), their *fate* in chimeras is to generate portions, which can be arbitrarily large, of a human body or brain. While the chimeric animals currently produced are "mostly nonhuman," the exact same techniques could lead to chimeric organisms that are "mostly human." This reality, which would be disturbing to large segments of the public, for reasons having nothing to do with rejection of abortion rights, is not generally appreciated, and the interested scientists and bioethicists seem set on keeping it that way. The availability of iPS cells has also mooted proposals to use molecular biological means to circumvent the use of embryos to obtain ES cells (Hurlbut 2005b; but see Byrnes 2010).

A concerted lobbying campaign by members of the stem cell research community (Sharma et al. 2015) claiming, with little concrete evidence, that "tremendous potential exists to develop humanized disease models for studying drug pharmacology . . . illuminate genetic disease pathogeneses . . . [and] to generate an unlimited supply of therapeutic replacement organs," finally led the NIH to throw in the towel. In the summer of 2016, the NIH issued an advisory lifting the ban on funding human-animal chimera research (Kaiser 2016). In fewer than 20 years and without public debate, research once considered "outlandish" and beyond the pale by prestigious members of the scientific community now qualified for public funding.

Conclusion: The Future of Human Developmental Modification

The impulse to genetically alter one's offspring arises from the desire to make them biologically "better" than they would otherwise have been. What, then, will be the attitude toward the unsuccessful products of these inherently fallible methods, including the nuclear transfer protocols approved for mitochondrial disease? Since some attempts will undoubtedly appear to be improvements, we might anticipate the rise of an Apple Inc.-type ethos for human production, with design or manufacturing errors becoming less and less acceptable to consumers. Under these circumstances, will adoption agencies then become routes for the placement of "factory seconds"?

It can be seen from the foregoing that the entry of different aspects and versions of the human organism into the "circulation of commodities" (Marx, Fowkes, and Fernbach 1981, Ch. 3) represents a sea change in our civilization. True to the hegemonic grip of progress and the marketplace, scholars, governmental panelists, journalists, and pundits have frequently dealt with these efforts by false analogies, failure to take account of new science relevant to the technologies, and occasionally outright deception. In other cases, where analysts have made a serious attempt to deal with philosophical issues raised by the construction of quasi-human animals, this shared ideology has led them to focus on questions such as the moral status of interspecies beings (Robert and Baylis 2003), rather than the broader impact on a culture in which the "human" has become a fluid or fungible quantity.

Regarding "intended humans," the only way to avoid problems arising from technological mishaps is not to modify human embryos at all.[4] Given the lines of force in contemporary society, however, this appears to be an unlikely prospect. Unwilling or unable to recognize challenges to received concepts in biology, no university-based reproductive bioethicists, whatever concerns they express about the equitable distribution of the anticipated benefits of gene manipulation, have publicly expressed doubts about the ability of the program to deliver on its promises. Most opponents of human genetic modification are similarly credulous on this issue (see, for example, Comfort 2015; Phillippidis 2016). Genetic determinism, however erroneous, is such a compelling ideology, and the reality of developmental processes so complex and multi-causal, that there is little chance that analysts and regulators will appreciate the recklessness of the effort before it is well underway.

Then there is the prospect that the "bad" outcomes will actually find uses—in research laboratories where they will be used to study developmental processes, in organ donation clinics, and in pharmaceutical companies seeking humanoid experimental models on which to test new drugs. Once we begin thinking of human-type organisms not as anybody's children or parents, a whole new world opens up for biotechnology to exploit. In the words of Drew Endy, a Stanford University bioengineer:

If you look at human beings as we are today, one would have to ask how much of our own design is constrained by the fact that we have to be able to repro-duce. . . . If you could complement evolution with a secondary path, decode a genome, take it off-line to the level of information . . . we can then design whatever we want, and recompile it. . . . At that point, you can make dispos-able biological systems that don't have to produce offspring.

(Cited in Specter 2009)

With the objective thus being "meiogenics" (from the Greek μείον: less), a kind of inverse of eugenics ("better inheritance"), many barriers fall aside (Newman 2012). Existing regulatory regimes on human experimentation pertain to what are agreed-upon humans; other, more permissive experimental regimes, cover verte-brate animals. If synthetic biologists can calibrate and titrate biological human-ity and its characteristic consciousness by taking the human genome offline and recompiling it, in 20 years we may be faced with all manner of humanoid organ-isms, serving various practical purposes—"lemonade" salvaged from the lemons of transhumanist experimentation.

It is not clear who will make the cut of being human, who will not, and who will decide (Newman 2003). What is clear is that the new reproductive technologies are blurring the theoretical and, in some cases, the practical, boundaries between unprecedented twins and templated clones, humans and nonhuman animals, and humans and useful humanoids. Without a real reversal in conventional think-ing about these questions and the institution of a taboo on modifying human embryos, it is difficult to see how we will not ultimately be led to the cultural establishment and legal enforcement of new boundaries among various categories of human hits and misses.

Acknowledgments

I thank the following individuals for their generous advice on content and rhetoric during the preparation of this essay: Diane Beeson, Irus Braverman, Malcolm Byrnes, James Siegel, and Tina Stevens.

Notes

1 Indeed, this is a current research paradigm (Hosseini et al. 2015). It is interesting to note in this regard that major morphological features, such as the number of vertebrae, can be determined by the egg cytoplasm rather than the nuclear genome in cross-species nuclear transfer experiments (Sun et al. 2005).

2 The Food and Drug Administration (FDA) issues guidelines to organizations engaged in research and development involving genetically modified organisms. Unlike its jurisdiction over food use of such organisms, the FDA's role in in their production is purely advisory. According to FDA Guidance for Industry Document #187, "FDA's guidance documents do not establish legally enforceable responsibilities . . . The use of the word 'should' in Agency's guidances means that something is suggested or recommended, but not required" (2015, 5).

3 The novel use of the patent process to serve activist or non-commercial ends attracted strong criticism. After two decades, however, the strategy has come to be deployed as part of a campaign to gain acceptance for a radical technology (see Braverman, Introduction to this volume).
4 It should not require stating that a proposed ban on modifying human embryos has no bearing on a woman's right to terminate a pregnancy.

References

Alliance for Humane Biotechnology. n.d. Available at: www.humanebiotech.org
Almeida-Porada, Graça et al. 2007. "The Human–Sheep Chimeras as a Model for Human Stem Cell Mobilization and Evaluation of Hematopoietic Grafts' Potential." *Experimental Hematology* 35: 1594–1600.
Americans for Cures. n.d. "Our Mission." Available at: http://americansforcures.org/our-mission
Annas, George J., Lori B. Andrews, and Rosario M. Isasi. 2002. "Protecting the Endangered Human: Toward an International Treaty Prohibiting Cloning and Inheritable Alterations." *American Journal of Law and Medicine* 28: 151–178.
Bahadur, Gulam et al. 2008. "Admixed Human Embryos and Stem Cells: Legislative, Ethical and Scientific Advances." *Reproductive Biomedicine Online* 17 (Supplement 1): 25–32.
Baltimore, David et al. 2015. "A Prudent Path Forward for Genomic Engineering and Germline Gene Modification." *Science* 348: 36–38.
Baylis, F. 2013. "The Ethics of Creating Children with Three Genetic Parents." *Reproductive BioMedicine Online* 26 (6): 531–534. DOI: 10.1016/j.rbmo.2013.03.006
BBC America. n.d.. *Orphan Black*. Available at: www.bbcamerica.com/shows/orphan-black
Begley, Sharon. 2016. "Do CRISPR Enthusiasts Have Their Head in the Sand About the Safety of Gene Editing?" *STAT Reporting From the Frontiers of Health and Medicine*. Available at: https://www.statnews.com/2016/07/18/crispr-off-target-effects
Billings, Paul R., Ruth Hubbard, and Stuart A. Newman. 1999. "Human Germline Gene Modification: A Dissent." *Lancet* 353: 1873–1875.
The Boys from Brazil. 1978. Motion Picture. Directed by Franklin Schaffner. Santa Monica, CA: Artisan Home Entertainment. (Video recording 1999.)
Briggs, Robert and Thomas J. King. 1952. "Transplantation of Living Nuclei from Blastula Cells into Enucleated Frogs' Eggs." *Proceedings of the National Academy of Sciences* 38: 455–463.
Brokowski, Carolyn, Marya Pollack, and Robert Pollack. 2015. "Cutting Eugenics Out of CRISPR-Cas9." *Ethics in Biology, Engineering and Medicine* 6: 263–279.
Byrnes, Walton M. 2005. "Why Human 'Altered Nuclear Transfer' is Unethical: A Holistic Systems View." *National Catholic Bioethics Quarterly* 5: 271–279.
———. 2010. "A Biomedical Revolution. The Pro-Life Promise of a New Stem Cell Technology." *America Magazine: The Jesuit Review*, August 16: 16–18.
Callaway, E. 2014. "Reproductive Medicine: The Power of Three." *Nature* 509 (7501): 414–417. DOI: 10.1038/509414a
Chinnery, P.F., L. Craven, S. Mitalipov, J.B. Stewart, M. Herbert, and D.M. Turnbull. 2014. "The Challenges of Mitochondrial Replacement." *PLoS Genetics* 10 (4): e1004315. DOI: 10.1371/journal.pgen.1004315
Christian, Lisa M. et al. 2015. "Gut Microbiome Composition is Associated with Temperament During Early Childhood." *Brain, Behavior, and Immunity* 45: 118–127.

Church, George M. and Edward Regis. 2012. *Regenesis: How Synthetic Biology Will Reinvent Nature and Ourselves.* New York: Basic Books.

CIRM. n.d. "Our Mission." California Institute for Regenerative Medicine. Available at: https://www.cirm.ca.gov/about-cirm/our-mission

Comfort, Nathaniel. 2015. "Can We Cure Genetic Diseases Without Slipping into Eugenics?" *The Nation,* July 15.

Connor, Steve. 2015. "Scientist Who Pioneered 'Three-Parent' IVF Embryo Technique Now Wants to Offer it to Older Women Trying for a Baby." *The Independent.* February 7.

Coughlin, Sean M. 2006. "The Newman Application and the USPTO's Unnecessary Response." *Chicago-Kent Journal of Intellectual Property* 5: 90–105.

Darnovsky, Marcy. 2008. "Germline Modification Carries Risk of Major Social Harm." *Nature* 453: 720.

Dawkins, Richard. 1998. "What's Wrong with Cloning?" In *Clones and Clones: Facts and Fantasies About Human Cloning.* Edited by Martha C. Nussbaum and Cass R. Sunstein, 54–66. New York: Norton.

Dickenson, D.L. 2013. "The Commercialization of Human Eggs in Mitochondrial Replacement Research." *The New Bioethics* 19 (1): 18–29.

Dowie, Mark. 2004. "Gods and Monsters." *Mother Jones* (January–February). Available at: www.motherjones.com/politics/2004/01/gods-and-monsters

Elliott, Carl. 2001. "Pharma Buys a Conscience." *The American Prospect* 12: 16–20.

Evans, Nicholas G. and Jonathan D. Moreno. 2015. "Children of Capital: Eugenics in the World of Private Biotechnology." *Ethics in Biology, Engineering and Medicine* 6: 285–297.

Farahany, Nita. 2014. "FDA Considers Controversial Fertility Procedure. What's at Stake?" *The Volokh Conspiracy,* February 25. Available at: https://www.washingtonpost.com/news/volokh-conspiracy/wp/2014/02/25/fda-considers-controversial-fertility-procedure-whats-at-stake/?utm_term=.89395b538699

FDA. 2015. *Guidance for Industry: Regulation of Genetically Engineered Animals Containing Heritable Recombinant DNA Constructs.* Available at: www.fda.gov/downloads/AnimalVeterinary/GuidanceComplianceEnforcement/GuidanceforIndustry/ucm113903.pdf

Fehilly, Carole B., S.M. Willadsen, and Elizabeth M. Tucker. 1984. "Interspecific Chimaerism Between Sheep and Goat." *Nature* 307: 634–636.

Fulka, Josef, Jr. and Helena Fulka. 2007. "Somatic Cell Nuclear Transfer (SCNT) in Mammals: The Cytoplast and its Reprogramming Activities." *Advances in Experimental Medicine and Biology* 591: 93–102.

Genetic Literacy Project. n.d. Available at: https://www.geneticliteracyproject.org

George, Robert P. 2016. *Conscience and its Enemies: Confronting the Dogmas of Liberal Secularism.* Wilmington, DE: Intercollegiate Studies Institute.

Gershoni, Moran et al. 2014. "Disrupting Mitochondrial-Nuclear Coevolution Affects OXPHOS Complex I Integrity and Impacts Human Health." *Genome Biology and Evolution* 6: 2665–2680.

Gould, Stephen J. 1997. "Individuality: Cloning and the Discomfiting Cases of Siamese Twins." *The Sciences* 37 (July–August): 14–16.

———. 1998. "Dolly's Fashion and Louis's Passion." In *Clones and Clones: Facts and Fantasies About Human Cloning.* Edited by Martha C. Nussbaum and Cass R. Sunstein, 41–53. New York: Norton.

Greely, Henry T. 2016. *The End of Sex and the Future of Human Reproduction.* Cambridge, MA: Harvard University Press.

Gurdon, J.B. 1968. "Transplanted Nuclei and Cell Differentiation." *Scientific American* 219: 24–35.

Habermas, Jürgen. 2003. *The Future of Human Nature.* Cambridge: Polity.

Hamzelou, Jessica. 2016. "Exclusive: World's First Baby Born With New '3 Parent' Technique." *New Scientist.* Available at: https://www.newscientist.com/article/2107219-exclusive-worlds-first-baby-born-with-new-3-parent-technique

Heled, Yaniv. 2014. "On Patenting Human Organisms or How the Abortion Wars Feed into the Ownership Fallacy." *Cardozo Law Review* 36: 241–298.

HFEA. 2016. "Code and Practice: Research and Training." Human Fertilisation and Embryology Authority. Available at: www.hfea.gov.uk/3468.html

Hosseini, S.M. et al. 2015. "Developmental Competence of Ovine Oocytes After Vitrification: Differential Effects of Vitrification Steps, Embryo Production Methods, and Parental Origin of Pronuclei." *Theriogenology* 83: 366–376.

Hurlbut, William B. 2005a. "Altered Nuclear Transfer as a Morally Acceptable Means for the Procurement of Human Embryonic Stem Cells." *National Catholic Bioethics Quarterly* 5: 145–151.

———. 2005b. "Altered Nuclear Transfer: A Way Forward for Embryonic Stem Cell Research. *Stem Cell Review* 1: 293–300.

Huxley, Aldous. 1932. *Brave New World, A Novel.* Garden City, NY: Doubleday Doran.

Hyun, Insoo, Amy Wilkerson and Josephine Johnston. 2016. "Embryology Policy: Revisit the 14-Day Rule." *Nature* 533: 169–171.

Ishii, T. 2014. "Potential Impact of Human Mitochondrial Replacement on Global Policy Regarding Germline Gene Modification." *Reproductive BioMedicine Online* 29 (2): 150–155.

James, Daylon, Scott A. Noggle, Thomasz Swigut, and Ali H. Brivanlou. 2006. "Contribution of Human Embryonic Stem Cells to Mouse Blastocysts." *Developmental Biology* 295: 90–102.

Kaiser, Jocelyn. 2016. "NIH Plans to Fund Human-Animal Chimera Research." *Science* 353: 634–635.

Krimsky, Sheldon and Jeremy Gruber. 2013. *Genetic Explanations: Sense and Nonsense.* Cambridge, MA: Harvard University Press.

Lanphier, Edward et al. 2015. "Don't Edit the Human Germ Line." *Nature* 519: 410–411.

Leonard, Kimberly. 2016. "What Ever Happened to Cloning?" *U.S. News.* August 4.

Levin, Ira. 1976. *The Boys from Brazil: A Novel.* New York: Random House.

Lippman, Abby et al. 1993. "Position Paper on Human Germ Line Manipulation Presented by Council for Responsible Genetics, Human Genetics Committee Fall, 1992." *Human Gene Therapy* 4: 35–37.

Liskovykh, Mikhail et al. 2015. "Stable Maintenance of De Novo Assembled Human Artificial Chromosomes in Embryonic Stem Cells and their Differentiated Progeny in Mice." *Cell Cycle* 14: 1268–1273.

Longaker, Michael T., Laurence C. Baker, and Henry T. Greely. 2007. "Proposition 71 and CIRM—Assessing the Return on Investment." *Nature Biotechnology* 25: 513–521.

Magnani, Thomas A. 1999. "The Patentability of Human-Animal Chimeras." *Berkeley Technology Law Journal* 14: 443–460.

Marx, Karl, Ben Fowkes, and David Fernbach. 1981. *Capital: A Critique of Political Economy, v. 1.* New York: Penguin Books in association with New Left Review.

Meinecke-Tillmann, Sabine and Burkhard Meinecke. 1984. "Experimental Chimaeras—Removal of Reproductive Barrier Between Sheep and Goat." *Nature* 307: 637–638.

Myhrvold, N. 1997. "Human Clones: Why Not?" *Slate.* Available at: www.slate.com/articles/briefing/critical_mass/1997/03/human_clones_why_not.html

Narasimhan, Vagheesh M. et al. 2016. "Health and Population Effects of Rare Gene Knockouts in Adult Humans with Related Parents." *Science* 352: 474–477.

Newman, Stuart A. 2003. "Averting the Clone Age: Prospects and Perils of Human Developmental Gene Manipulation." *Journal of Contemporary Health Law and Policy* 19: 431–463.

———. 2006. "My Attempt to Patent a Human-Animal Chimera." *L'observatoire de la génétique* 27 (April–May).

———. 2012. "Meiogenics: Synthetic Biology Meets Transhumanism." *GeneWatch* 25: 31–37.

NIH. 2009. "National Institutes of Health Guidelines on Human Stem Cell Research." Available at: https://stemcells.nih.gov/policy/2009-guidelines.htm

———. 2015. "NIH Research Involving Introduction of Human Pluripotent Cells into Non-Human Vertebrate Animal Pre-Gastrulation Embryos." Available at: https://grants.nih.gov/grants/guide/notice-files/NOT-OD-15-158.html

Pacholczyk, Tadeusz and J. Benjamin Hurlbut. 2005. "The Substantive Issues Raised by Altered Nuclear Transfer." *National Catholic Bioethics Quarterly* 5: 17–19, 19–22.

Palacios-Gonzalez, C. 2016. "Mitochondrial Replacement Techniques: Egg Donation, Genealogy and Eugenics." *Monash Bioethics Review.* DOI: 10.1007/s40592-016-0059-x

Pence, Gregory E. 1998. *Who's Afraid of Human Cloning?* Lanham, MD: Rowman & Littlefield.

———. 2004. *Cloning after Dolly: Who's Still Afraid?* Lanham, MD: Rowman & Littlefield.

———. 2012. *How to Build a Better Human: An Ethical Blueprint.* Lanham, MD: Rowman & Littlefield.

———. 2016. *What We Talk About When We Talk About Clone Club: Bioethics and Philosophy in Orphan Black.* Dallas, TX: Smart Pop Books.

Pereira, Carlos F. et al. 2008. "Heterokaryon-Based Reprogramming of Human B Lymphocytes for Pluripotency Requires Oct4 but not Sox2." *PLoS Genetics* 4: e1000170.

Peura, Teija T., Julie Schaft, and T. Stojanov. 2010. "Derivation of Human Embryonic Stem Cell Lines from Vitrified Human Embryos." *Methods in Molecular Biology* 584: 21–54.

Phillippidis, Alex. 2016. "Pursuing CRISPR Vision on Germline Editing." *Clinical OMICs* 3: 17–19.

Philpott, Tom. 2012. "The Making of an Agribusiness Apologist." *Mother Jones.* February 24.

Pinker, Steven. 2015. "The Moral Imperative for Bioethics." *Boston Globe.* August 1. Available at: https://www.bostonglobe.com/opinion/2015/07/31/the-moral-imperative-for-bioethics/JmEkoyzlTAu9oQV76JrK9N/story.html

Reardon, Sara. 2016. "US Panel Greenlights Creation of Male 'Three-Person' Embryos." *Nature* 142. Available at: www.nature.com/news/us-panel-greenlights-creation-of-male-three-person-embryos-1.19290

Ridley, Matt. 2015. "Mitochondrial Donation is a Wonderful Opportunity." The Rational Optimist, February 2. Available at: www.rationaloptimist.com/blog/mitochondrial-donation-is-a-wonderful-opportunity

Robert, Jason and Francoise Baylis. 2003. "Crossing Species Boundaries." *American Journal of Bioethics* 3: 1–13.

Ruskin, Gary. 2015. "A Short Report on Jurnalists Mentioned in our FOIA Requests." *U.S. Right to Know.* September 28. Available at: http://usrtk.org/gmo/a-short-report-on-journalists-mentioned-in-our-foias

Sharma, Arun et al. 2015. "Lift NIH Restrictions on Chimera Research." *Science* 350: 640.

Sinclair, Kevin D. et al. 2016. "Healthy Ageing of Cloned Sheep." *Nature Communications* 7: 12359.

Singer, Peter. 2003. "Shopping at the Genetic Supermarket." In *Asian Bioethics in the 21st Century*. Edited by Sang-yong Song, Young-Mo Koo, and Darryl R. J. Macer. Christchurch, NZ: Eubios Ethics Institute.

Sleeper. 1973. Motion Picture. Directed by Woody Allen. US: United Artists Corporation and MGM Home Entertainment Inc.

Specter, Michael. 2009. "A Life of Its Own." *The New Yorker*. September 28.

Staff, Science News. 1997. "Harkin on Human Cloning." *Science*. Available at: www.sciencemag.org/news/1997/03/harkin-human-cloning-i-welcome-it

STAT-Harvard. 2016. *The Public and Genetic Editing, Testing, and Therapy*. Available at: https://cdn1.sph.harvard.edu/wp-content/uploads/sites/94/2016/01/STAT-Harvard-Poll-Jan-2016-Genetic-Technology.pdf

Stein, R. 2016. "In Search for Cures, Scientists Create Embryos That Are Both Animal and Human." In *All Things Considered*. USA: National Public Radio.

Stevens, M. L. Tina. 2000. *Bioethics in America: Origins and Cultural Politics*. Baltimore, MD: Johns Hopkins University Press.

———. 2007. "Intellectual Capital and Voting Booth Bioethics." In *The Ethics of Bioethics: Mapping the Moral Landscape*. Edited by Lisa A. Eckenwiler and Felicia G. Cohn, 59–73. Baltimore, MD: Johns Hopkins University Press.

Subbaraman, Nidhi. 2016. "No One Should Edit the Genes of Embryos to Make Babies, NIH Chief Says." *BuzzFeedNEWS*. July 14. Available at: https://www.buzzfeed.com/nidhisubbaraman/gene-editing-ethics?utm_term=.nvDn1zoKKW#.ih82OaPNNk

Sun, Yong-Hua et al. 2005. "Cytoplasmic Impact on Cross-Genus Cloned Fish Derived from Transgenic Common Carp (*Cyprinus carpio*) Nuclei and Goldfish (*Carassius auratus*) Enucleated Eggs." *Biology of Reproduction* 72: 510–515.

Trounson, Alan, Robert Klein, and Richard Murphy. 2008. "Stem Cell Research in California: The Game is On." *Cell* 132: 522–524.

True, John R. and Eric S. Haag. 2001. "Developmental System Drift and Flexibility in Evolutionary Trajectories." *Evolution and Development* 3: 109–119.

U.S. Government. 1949. "The Nuremberg Code." In *Trials of War Criminals before the Nuremberg Military Tribunals under Control Council Law No. 10*. Washington, D.C.: U.S. Government Printing Office.

U.S. Patent Application No. 08/993,564. 1997. "Chimeric Embryos and Animals Containing Human Cells." Filed December 18.

Vogel, Gretchen. 2015. "Mitochondrial Gene Therapy Passes Final U.K. Vote." *Science*, February 24. Available at: www.sciencemag.org/news/2015/02/mitochondrial-gene-therapy-passes-final-uk-vote

Weiss, Rick. 1999. "U.S. Ruling Aids Opponent of Patents for Life Forms." *Washington Post*. June 17.

———. 2005. "U.S. Denies Patent for a Too-Human Hybrid." *Washington Post*. February 13.

Wilmut, Ian et al. 1997. "Viable Offspring Derived From Fetal and Adult Mammalian Cells." *Nature* 385: 810–813.

WMA. 2013. "World Medical Association Declaration of Helsinki: Ethical Principles for Medical Research Involving Human Subjects." *JAMA* 310: 2191–2194.

Wolf, D.P., N. Mitalipov, and S. Mitalipov. 2015. "Mitochondrial Replacement Therapy in Reproductive Medicine." *Trends in Molecular Medicine* 21 (2): 68–76. DOI: 10.1016/j.molmed.2014.12.001

Yamanaka, Shinya and Kazutoshi Takahashi. 2006. "Induction of Pluripotent Stem Cells from Mouse Fibroblast Cultures." *Tanpakushitsu Kakusan Koso* 51: 2346–2351.

Zwerdling, Daniel. 1998. "Humanimals." In *All Things Considered*. USA: National Public Radio.

Chapter 8

Domestic Dogs, Gene Repair, and the "One Health" Approach

Alexander J. Travis

Dog Domestication and Breeds

People love dogs. Approximately 80 million dogs live as pets in the United States (American Society for the Prevention of Cruelty to Animals 2016). This love affair has been going on for some time. The dog (*Canis familiaris*) was the first species to be domesticated, some 18,000–32,000 years ago, long before humans domesticated agricultural animals and even before we cultivated crops (Thalmann et al. 2013, 872). These early dogs were an integral part of human existence, moving with humans in the late Pleistocene epoch to the New World over the Bering land bridge—the only domestic species to reach the Americas prior to oceanic travel (Leonard et al. 2002, 1613). Originating from the grey wolf (*Canis lupus*) (Lindblad-Toh et al. 2005, 815) in either one or two separate domestication events (Frantz et al. 2016), dogs themselves can be thought of as the result of the first experiments ever performed in genetic modification. This species would simply not exist if not for the interactions of wolves with humans and vice versa, that over generations led to dramatic changes in the appearance and behavior of both wolves and humans.

Human selection over this time has produced over 500 distinct breeds, created to carry out diverse functions such as hunting, herding, protection, guidance, and draft power (Schoenebeck and Ostrander 2014, 537). However, the greatest radiation of diversity has only occurred since the nineteenth century, as strict breed standards were developed (Larson et al. 2012). Selected for physical and behavioral traits, the variation among breeds in the single species of *Canis familiaris* is astonishing and dwarfs that in any other species. Using a well-known comparison, the breed standard for the Chihuahua is not to exceed 6 pounds, whereas mastiffs, Newfoundlands and St. Bernards can exceed 150 pounds, a 25-fold difference (American Kennel Club n.d.).

Genetic Linkage and Predisposition to Disease

The genetic modification of dogs has been central to the domestication process. For most dog breeds, there were two major periods in which "genetic bottlenecks" have occurred. The first and greatest loss of genetic diversity occurred with the initial domestication process. As breed standards tightened and specific traits/ genes desired exclusively, there has also been loss of genetic diversity in those areas of the genome, a phenomenon known as a "selective sweep" (Marsden et al. 2016, 155). To understand how positive genetic selection for desired traits can lead to increased risk of disease, I'll begin with a simple example of how various traits can be linked.

Modern experimental studies of generations of silver foxes bred for "tame" versus more "wild" behaviors have revealed a close association between genes and developmental processes involved in multiple aspects of behavior and those that shape appearance. Foxes bred for dog-like behavior soon began to look like dogs, with shortened muzzles, floppy ears, and mottled coat colors (Trut 1999, 163). In addition, without any training, these animals were shown to have reduced stress responses in proximity to humans, different vocalization patterns, and the ability to read human behavioral cues (ibid., see also Kukekova et al. 2012). These results can be explained by a phenomenon known as "genetic linkage." When genes are close together on a chromosome, they tend to be inherited together because they are less likely to be separated during chromosomal recombination as the developing gametes undergo meiosis. If there is a strong selection pressure for a certain gene (whether natural or artificial pressure based on human choice), then nearby genes come along for the ride.

Because breeds are defined by many characteristics, we usually do not know exactly which preferred traits might be linked to which deleterious genes, or whether those linkages are even between "breed defining" traits, or instead are associated with general features, such as size. However, in total, this phenomenon has resulted in profound, breed-specific genetic predisposition to disease, which from a health perspective can be thought of as defining characteristics in their own right (however undesirable). Just as we know that Dalmatians are white dogs with black spots, we also know that all Dalmatians carry a mutant form of the *SLC2A9* gene that leads to high levels of uric acid and predisposes them to urinary stones and blockage (Bannasch et al. 2008). Just as we know that golden retrievers have friendly dispositions and soft coats, we know that they are also predisposed to two forms of cancer, lymphoma and hemangiosarcoma (Tonomura et al. 2015). Scottish deerhounds, Irish wolfhounds, greyhounds and Rottweilers are predisposed to osteosarcoma (Schoenebeck and Ostrander 2014, 541); Doberman pinschers are predisposed to familial cardiomyopathy, resulting from a deletion of part of the gene encoding pyruvate dehydrogenase kinase 4 (Meurs et al. 2012, 1322). Lest one think that predisposition to genetic disease is limited to large breed dogs, West Highland white terriers are predisposed to food allergies (Scott et al. 2001, 618), and congenital patent ductus arteriosus, a disorder of cardiovascular development, has been identified in Chihuahuas (Bomassi et al. 2011). The list of breeds and the diseases to which they are predisposed seems endless. In fact, veterinarians and genomic scientists have to date identified 678 total traits and disorders in dogs, of which 383 are potential models for a homologous human disease (Nicholas n.d.).

"One Health": The Dog as a Biomedical and Genetic Model

Because of the anatomic and physiologic similarities to humans, and the genetic basis of many canine diseases, understanding the nature and treatment of pathologies in dogs has the potential to help when confronting the same conditions in humans. This approach is known as "comparative medicine," and is one of the pillars of the "One Health" paradigm, which teaches that the health of humans, nonhuman animals, and the environment is not just interconnected, but mutually dependent (American Veterinary Medical Association n.d.a). Although people have been studying disease in animals for the benefit of humans for centuries, the paradigm of One Health helps us appreciate other aspects of the almost unique relationship between domestic dogs and humans. For example, dogs share our household environments to an extent that is unmatched by any other species other than the cat, so, with their shorter lifespans, they can function as sentinels for health impacts that we ourselves face. Dogs have become an important part of our lives, conferring both important physical and psychological health benefits (Hart 1995). Because they are often thought of as members of the family, people are deeply interested in preserving the quality of their lives through the advancement of medical knowledge and therapies.

When discussing studies of genetic disease in dogs, it is important to distinguish this work from the approach commonly used with laboratory mice. In mice, when scientists wish to know what the product of a gene does inside a cell or tissue, they can use transgenic technology to delete or alter that gene. Scientists then determine the function of that gene by observing the resultant phenotype (i.e., what goes wrong in the genetically modified mouse?). As we've seen, the situation is the exact opposite in the dog—we have already performed the grand genetic modification experiment in having created 500 dog breeds, and we already know the results in the form of different behaviors, physical characteristics, and predispositions to disease. But unlike the mouse, we just don't know which gene or genes are responsible for the differences that we can observe. This unprecedented genetic experiment, underway for centuries, has created an incredibly rich opportunity for scientists to unlock the genetic basis of disease, morphological traits (such as the length of bones, or shape of the skull), and behaviors. Through the use of a variety of genetic tools and approaches, scientists have been able to identify the key mutations responsible for 206 of these genetic traits and disorders in dogs (Nicholas n.d.)

Using Genetic Knowledge to Advance Health

Knowing which gene or genes are responsible for a condition opens several pathways to prevent or reduce suffering in both people and dogs. First, knowledge of the defective gene in the dog or human can help scientists identify whether the same gene is responsible for the condition in the other species. In both cases, this can lead to development of a pharmaceutical therapy. Second, and more relevant to inquiries into questions of ethics and genetics, one can design genetic screening tests. In humans, the field of genetic counseling has arisen to help people at risk of passing on a genetic condition to make informed decisions. The technology of preimplantation genetic diagnosis (PGD) is used with assisted reproductive technologies in which embryos are grown in culture to provide precise genetic understanding about individual embryos. Single cells (called "blastomeres") are removed from the embryos and are screened to determine chromosomal gender in the case of sex-linked traits (Handyside et al. 1990), or to see whether they carry one or more deleterious traits (Sermon et al. 2004). This technique helps clinicians and couples determine which embryos they would like to use for embryo transfer. While these are important tools in human medicine, such screening approaches vary in value in dogs. Starting with the latter case of PGD, this technique is not used in veterinary medicine because *in vitro* fertilization (IVF) has just been performed in dogs for the first time (Nagashima et al. 2015; see Figure 6). Therefore, no dog embryos have been available for such evaluation. Even in the unlikely event that canine IVF becomes a widespread breeding tool, PGD is still unlikely to become a common procedure because of cost and because screening prior to breeding would be used to remove individual animals from the breeding pool and thereby obviate the need for PGD.

Figure 6 The world's first litter of puppies born by *in vitro* fertilization (IVF). Embryos were produced in the laboratory by IVF, stored frozen (cryopreserved), then thawed and transferred into a recipient female hound. She gave birth to these seven healthy puppies, which, in January 2017, were 18 months old. IVF and embryo cryopreservation are important tools to help make gene repair practical in dogs.

Courtesy of author and Cornell University.

Several companies have emerged that offer the genetic screening of dogs in order to assist breeders in making informed decisions. This approach can be very successful at preventing known carriers from being bred together in populations in which the deleterious trait has not become fixed in the population. For example, the estimated genotypic prevalence of Type 2 von Willebrand disease (a severe coagulopathy) in German wirehaired pointers is 8.2 percent (Gavazza et al. 2012, 1465). Genetic screening could likely be very efficacious in such circumstances at removing the mutant allele from the population. However, in situations such as the Dalmatian and its defect in *SLC2A9*, all members of the breed possess the same mutant allele, and selective breeding is not possible. In situations in which the mutant allele is common but not ubiquitous, one must evaluate whether selective breeding will pose a greater risk for the breed than the condition itself, as there could be substantial loss of genetic diversity within the limited population if the number of acceptable breeders drops too low. In other words, one could lose the predisposition to the known threat, but fix another deleterious trait.

Alternatively, one can design treatments for a disease that seek to rescue the defective gene by replacement of the gene or by manipulating the genome. In what is probably the best example, the genetic basis of a severe form of childhood blindness (Leber congenital amaurosis) was identified in both dogs and humans as resulting from mutations in *RPE65* (Acland et al. 2001, 92). This disease causes near-total blindness in infancy. Using an adenovirus as a vehicle to deliver the functioning gene to the retinal cells, researchers successfully performed gene therapy—restoring sight in the treated eyes of Briard dogs as demonstrated by electroretinography, pupillary response, and qualitative evaluation of sight and avoidance of obstacles (ibid., 94). This technique has subsequently been performed in humans, restoring modest sight to the blind (Bainbridge et al. 2008). Continued work in this area on both humans and dogs has revealed some difference in terms of the amount of virus needed, and suggests that the effect was not permanent in humans (Bainbridge et al. 2015, 1887). Earlier intervention and lasting expression of appropriate amounts of the gene product are likely required for humans to show the same degree of response as the dog (ibid., 1896). Replacement of a mutant allele with a functional copy at the start of life (a single-cell embryo) would solve these issues. But before discussing how gene editing and gene repair may be used, I would first like to address the underlying ethical concerns regarding dog breeding itself.

The Ethics of Dog Breeding

Concerns about the ethics of dog breeding typically take two independent, though related, forms. The first concern stems from the pet overpopulation crisis. Under these circumstances, is it justifiable to continue breeding dogs? This debate is focused on the welfare of the 3.9 million dogs that enter shelters in the United States each year—1.2 million of which are euthanized (American Society for the Prevention of Cruelty to Animals 2016). No one can deny the extent of this problem in the United States or internationally. Information campaigns that encourage the neutering of animals adopted from shelters have made important contributions. Yet clearly, much remains to be done. For many people who desire canine companionship in their household, mixed breed dogs obtained from animal shelters are the perfect solution. However, there are others who desire certain behaviors or morphological phenotypes. For example, apartment dwellers might need a small dog, families with young children might want a breed well recognized for being gentle companions, and hunters might need pointers or retrievers. Still others will need dogs specifically bred to perform highly specialized services such as police dogs, watchdogs, contraband-sniffing dogs, hypoallergenic dogs, or guide dogs for the blind or physically challenged. Although there will always be variation among individual dogs even within a breed, for people needing these specialized services, the advantages of the defined genetics offered by breeds are highly compelling.

The second ethical concern with dog breeding is whether it is ethical to perpetuate or accentuate traits that themselves cause distress or health problems for those animals. This debate concerns the application of breed standards, typically chosen to enhance a specific physical characteristic, which results in reduced fitness or welfare. The most common example cited in the literature is that of the English bulldog, which suffers from a variety of airway, breathing, and temperature regulation disorders due to its extremely foreshortened (brachycephalic) muzzle, and conformational changes affecting the palate and trachea (Denizet-Lewis 2011). This breed also has difficulties giving birth because of the width of the puppies' heads relative to the pelvic canal of the mother (ibid.). In this case, selection to meet changing aesthetic tastes has exacerbated these traits to the point where the dogs are no longer healthy.

These related issues are complex and have been discussed extensively (Horowitz et al. 2013). My personal ethics are informed by my training and by the oath that I have taken as a veterinarian:

> I solemnly swear to use my scientific knowledge and skills for the benefit of society through the protection of animal health and welfare, the prevention and relief of animal suffering, the conservation of animal resources, the promotion of public health, and the advancement of medical knowledge.
>
> (American Veterinary Medical Association n.d.b)

From my perspective as a scientist and practitioner, the genetic diversity represented by the different breeds is a tremendous resource, created over time, and can help promote the well-being of both humans and dogs in their varied relationships. Still, I cannot support the selection for traits that promote suffering and harm animal welfare. I would thus argue that breed standards that promote unhealthy traits should be changed. Gene repair is not needed here—the issue is choosing breeding pairs based on trying to accentuate an unhealthy trait versus trying to promote health, while retaining the other desired characteristics of that breed. What is needed here, then, is a regulatory standard that would be applied to breeders across the board, and that would emphasize health over aesthetics.

The Ethics of Gene Repair

Ethical debates aside, humans continue to breed dogs. These breeding efforts have resulted in well-recognized genetic predisposition of these dogs to disease. Given these conditions, I would argue that we owe it to these animals, who have become part of our families and lives and with whom we bond emotionally, to try to prevent their suffering.

As we left off above, the ability to replace a defective copy of a gene with a functional copy would be a powerful tool to prevent disease before it could occur. Gene editing technology provides a mechanism to achieve this goal. Combining components from the immune system of single-cell organisms, Clustered Regularly Interspaced Short Palindromic Repeats (CRISPR), together with a modified Cas9 nuclease, has led to the creation of the CRISPR-Cas9 system for gene editing (Jinek et al. 2012; Wiedenheft et al. 2012). Briefly, inclusion of a piece of RNA targets the nuclease Cas9 to a specific location in the genome, where it introduces a double-strand cut. Delivery of a repair template that contains the desired genetic change allows the cell's own DNA repair machinery to fix the cut based on the template (Singh et al. 2015; see also Esvelt, Chapter 1, this volume).

CRISPR-Cas9 potentially allows scientists to elegantly repair mutations or deleterious sequences in an allele with utmost precision. This approach is referred to as "gene repair," using a term from gene therapy that also involved precise replacement of a defective portion of a gene using homologous recombination (Parekh-Olmedo et al. 2005). In this context, the definition of gene repair refers exclusively to the replacement of the defective gene with a functional copy that originates from other individuals within that same species. Therefore, although the incorporation of a gene from another species and the deletion of a gene are both examples of gene editing, neither are examples of gene repair.

I have thus reached the central question of this chapter: is it ethical to perform gene repair in dogs using CRISPR-Cas9 technology? Since most people would likely agree that the prevention of suffering and the promotion of welfare in nonhuman animals are worthwhile goals, I will focus here on the ethics of the CRISPR technology itself. For this purpose, I will consider four central questions: Do we really need this technology? Does it cause suffering? Does it provide benefits? And does this technology cause harm beyond the genetically modified dogs? In what follows, I will address these questions in order.

First, do we really need to apply CRISPR-Cas9 to dogs? Having described this amazing new genetic tool, the first question to ask is whether traditional dog breeding approaches could solve problems without this technology. One might argue that gene repair using CRISPR-Cas9 is not ethical because it is not needed: we can perform out-crossing to introduce functional genes into the population, an approach that has indeed been used successfully in the Dalmatian to overcome the defect in *SLC2A9*. A pointer, a closely related breed with a functional *SLC2A9* gene, was bred just once to a Dalmatian. Offspring having low uric acid levels were then bred back to purebred Dalmatians, with the selection of subsequent offspring for low uric acid levels as well as breed-appropriate conformation and physical characteristics (Schaible 1981). After multiple generations of backcrossing (always to Dalmatians), it is impossible to visibly distinguish these Dalmatians from those who were not part of the project (ibid.). Despite the apparent success of this approach, its reception has been mixed (Bernardi 2010). Some have raised the concern that because of genetic linkage, other traits that lie near the *SLC2A9* gene on the chromosome would also be perpetuated—making these animals "less" of a Dalmatian. This is despite the fact that after more than 40 years of breeding since the single breeding with the pointer, these dogs are over 99 percent pure Dalmatian (ibid.). Other concerns revolve around the fact that the backcrossing process restricts the gene pool within the breed, which risks fixing other deleterious traits in the population. Clearly, there is disagreement among the individuals who care most for this breed.

From a practical perspective, this sort of breeding approach can take years to achieve. Implicit in these years are numerous generations of dogs that either still carry the mutant allele or have "unacceptable" amounts of mixed genetics ("unacceptable" is in quotation marks here because breed experts determine what is an acceptable level, and such a determination varies among breeds and changes over time). Ensuring the adoption of these numerous animals is not trivial, yet it is essential in avoiding an ethical issue inherent to generating potentially unwanted animals. A technique that can make a precise change quickly, within a single generation and with fewer animals, would be attractive in comparison.

The second question I would now like to address is whether CRISPR-Cas9 causes the dogs any suffering. Even if the goal of pinpoint alteration of a gene to prevent disease is seen as beneficial, one must balance that against the cost in terms of suffering and the number of animals needed to achieve that goal. Here, I will consider the technological barriers that need to be overcome for gene repair to work. Before beginning, one must remember that use of gene editing technology is in its infancy, with much information still needed for the approach to be used successfully in various species. Success rates for the first reports should thus be considered a starting place, as future studies will lead to optimized protocols.

Although gene repair in dogs has not yet been accomplished, the modification of the dog genome using CRISPR-Cas9 has already been reported (Zou et al. 2015). Briefly, the investigators targeted Cas9 to the beginning of the *myostatin (MSTN)* gene, which is a negative regulator of muscle growth (McPherron and Lee 1997). Loss of *MSTN* function in agricultural animals and mice results in excessive muscle development (known as a "double muscle" phenotype) (ibid. 12457). In this study in the dog, the goal was for the Cas9 nuclease to induce a double-stranded break that would be repaired by imprecise (non-homologous) end joining. Insertion or deletion of nucleotide bases early in a gene typically causes a frameshift (breaking the triplet pattern of reading bases) and ultimately results in a "knockout," or loss of function of the resulting protein. These investigators were successful: from 35 embryos that they injected, they produced one puppy that had bi-allelic loss of *MSTN*, and one puppy that was a chimera (some tissues being normal and some missing a functional *MSTN*) (Zou et al. 2015).

In this study, the investigators used CRISPR-Cas9 not to perform gene repair, but rather to perform gene editing—simply disrupting the function of *MSTN*. Even though this is technically less challenging than gene repair, there are still many steps that need to occur to get the desired result. First, the Cas9 must be targeted to the correct sequence. The Cas9 must then cut both strands of DNA at the right site and have them both be repaired in ways that introduce a frameshift. For many genes, alteration of one allele alone will not produce a change in function. During embryonic development, each cell of the embryo must inherit this change, or else it will develop as a chimera, with some cells with and some without the alteration. Lastly, the CRISPR-Cas9 should not bind off target sites on the genome, or else unwanted changes might occur. In this case, they obtained one puppy with the desired result (~3 percent of the injected one-cell embryos) (ibid.). In actuality, this overall yield was approximately one half of the potential yield because of the experimental design. In it, single-cell embryos were flushed from one oviduct, injected with the construct, and then transferred back into the oviduct on the other side of the same female (ibid.). This strategy was utilized because of difficulties in timing and managing canine reproduction, but one can see that the yield from the start is half of what it could be, because any embryos in the oviduct on the side that was not flushed would be lost from the procedure (ibid.).

Although a very respectable yield for a first-ever report, such a modest efficiency reflects the fact that dog reproduction is different from that of other mammals in many ways. For example, dogs ovulate a primary oocyte that is very immature relative to oocytes of other mammals such as humans (Reynaud et al. 2005). This means that more maturation must take place in the oviduct, making timing difficult. The oocytes and cells of the early embryo are almost black in color, due to a high lipid content (Chastant-Maillard et al. 2010, 1052), limiting the ability of scientists to see where they are injecting the CRISPR-Cas9 constructs, and potentially influencing the likelihood of obtaining the alteration in all or only some of the cells. Combine this with the facts that dogs only cycle once or twice a year and that we do not have highly effective synchronization procedures to regulate the timing of cycling (Concannon 2011, 209), and one can appreciate the difficulties these investigators faced.

A recent advancement has been made, however, with the potential to double the efficiency versus this initial report: my laboratory succeeded in performing IVF for the first time in a dog (Nagashima et al. 2015). This technique gives us the ability to generate and manipulate canine embryos in the laboratory. Oocytes can be collected from both sides of the female tract, as opposed to only one side, thereby doubling yield. And in this report, the early embryos were cryopreserved until a recipient could be timed appropriately for embryo transfer (ibid.). These new capabilities certainly do not address all the challenges imposed by performing gene editing or repair in the dog, but they provide a valuable start.

Performance of gene repair in dogs using CRISPR-Cas9 is still in its experimental stage. If the reported success rate of 3 percent for bi-allelic, germline alterations could be doubled or improved upon further, then it is likely that the number of animals needed to remove the gene from the breeding population would be lower than in the backcross approach, because one wouldn't have to breed multiple generations to obtain a genetically modified individual with a pure, breed-specific genetic background. In other words, CRISPR would allow scientists to do what dog breeders have been doing for centuries, except with more accuracy and efficiency and with reduced numbers of animals involved.

Does CRISPR-Cas9 provide benefits? Being able to repair genes that predispose an individual to disease would provide immeasurable improvement in those individuals' quality of life. Note that this is not a technique that would be performed once or even twice. If only one genetically altered male and one female were to be generated, and the breed repopulated from them, the loss of genetic diversity in the breed would be staggering. Rather, breeders who want to improve the genetics of their lines would have the procedure performed so that their animals could be used for future pairings. This is similar to the situation of trying to replace the deleterious gene by backcrossing. To avoid an extreme population bottleneck, one would have to introduce multiple lines of the breed into the program. Although the end result would be the same as the backcross approach, there is another potential benefit: time. The repaired gene could be introduced much more quickly throughout the population than would be possible if having to wait for multiple generations of dog to be produced.

There is an additional benefit to using CRISPR. Much as retinal gene therapy in the dog paved the way for trials in human patients, gene repair using canine embryos might provide valuable information on eventual use of the technique in humans. That possibility certainly raises another host of ethical concerns and is beyond the scope of this chapter. Yet to have an informed discussion in the future, we must know whether key scientific and technical challenges can be overcome and, if so, how. Research on optimization of the length of single guide RNA is already underway and advancements and refinements have already commenced (Singh et al. 2015). One simply cannot determine the feasibility of this approach without research to test and optimize it. The use of CRISPR-Cas9 in the dog might not only prevent disease in dogs, but also pave the way for improvements in human health.

Should it be deemed that gene repair is *not* appropriate for use in people, or that technical challenges make it too risky, there are still benefits to human health from pursuit of these studies in dogs. Namely, genetic studies can identify genes as strong candidates for causing a disease or influencing a trait. Yet in the absence of genomic modification, one cannot completely rule out (or understand the extent of) the other genes that contribute to that condition. That is, one might identify a certain nucleotide change in gene X as being likely to cause a disease, but until one can precisely repair that change, and that change only, it is difficult to completely rule out the possibility that gene Y might also play an important role. This is a significant benefit of introducing changes in the mouse genome to remove functions. One can identify where other gene products can help compensate for the loss of a given gene, or identify multiple changes that might be needed for a disease to manifest itself clinically. In the case of gene repair in the dog, one can determine whether successful replacement of the sequence believed to be responsible for the unwanted trait actually prevents the problem from occurring. If so, does it completely solve the problem in all individuals, or does the degree of improvement vary among individuals? If the latter, then other genes or environmental factors might play a role. This information could inform the use or design of gene therapy approaches as well as traditional drug development.

Fourth and finally is the question of whether the CRISPR technology can cause harm beyond the genetically modified dogs themselves. Unlike situations in which genetically modified organisms would be released into the environment to change ecological systems, *Canis familiaris* is a domestic species whose reproduction is largely controlled by humans. Certainly, the feral and stray dog population crisis informs us that we have been inadequate in those efforts at control. Yet it is highly unlikely that repair of genetic disease in a domestic animal would cause an ecological or public health crisis.

Together, gene repair using CRISPR-Cas9 offers several advantages to currently available backcrossing approaches, although one must be aware that gene repair is still at a highly experimental stage of development. In addition to potentially preventing animal suffering, research on gene repair could provide important information on the genetic nature and potential treatments of homologous conditions in humans. This is one of the tenets of the One Health paradigm, which stresses the interconnectedness of all life.

The Ethics of Gene Editing in Dogs Beyond Repair

As shown by the experiments with perturbation of *MSTN*, one can use CRISPR-Cas9 to do more than repair defective genes and prevent disease. Indeed, scientists can just as easily use the technology to knock out gene function and potentially induce disease. Additionally, scientists could use the technology to insert genes from other species, conferring the target species with new characteristics and abilities. One of the most likely uses for that approach would be to introduce a fluorescent marker protein, such as green fluorescent protein, into a specific cell or tissue.

Because of the incredible power of dogs as biomedical models for treating human conditions, the introduction of fluorescent marker protein has already been executed in dogs (Kim et al. 2011). Due to animal welfare concerns, as well as practical matters such as longer generation time, larger size and higher expenses, the dog will never replace the mouse as a laboratory animal. Nonetheless, dogs are preferred over rodents for some studies. In particular, their longer lifespan has important implications for hematopoietic, cancer, gene therapy, and stem cell-based studies. For example, dogs have been an ideal model for hematological transplant studies because the small size and short lifespan of rodents impart comparatively little proliferative demand on their hematopoietic stem cell and progenitor cell compartments. Put more simply, during its lifetime, a mouse makes as many red blood cells as a human makes in one day, and a dog makes in 2.5 days (Georges et al. 1998). These differences have made it difficult to extrapolate gene transfer and transplantation results from rodents to humans, whereas results from studies done in dogs have been shown to be more predictive of outcomes in human patients (Lupu and Storb 2007). This line of discussion raises a host of ethical questions regarding whether *any* animal-based research is appropriate for a study about humans, a debate that reaches far beyond the scope of this chapter.

Relatedly, ethical questions exist regarding the introduction of genes from certain species to others. Creating dogs with marked cells would provide great advantage in studies to develop new bone marrow or stem cell transplantation studies, helping both human and veterinary medical advancement and ultimately relieving suffering. I would like to distinguish this type of gene introduction from other species versus the creation of what up until now could only be considered as imaginary animals. In my view, it is unethical to mix and match genes from various species in attempts to create fantastical creatures. I do not wish for a world with zebra-striped dogs, or dogs with wings, or dogs with the trunks of elephants. Gaining such complex attributes would require the alteration of multiple genes and a complex interplay of multiple developmental pathways. The suffering that would be required to get to those end-points, in terms of the "failed experiments" along the way, would be astronomical. But more to the point, domestic dogs represent refinement of traits that have existed in the gene pool handed down from the original wolves, alongside those that have arisen from spontaneous mutation along the way. Mixing genes from another species into a dog would render it something different, and no longer a dog.

Gene Editing in the Conservation of Wild Species

Silently but pervasively, life on earth is disappearing. We are in the midst of the sixth global mass extinction crisis, which is biological in origin, with humans being the culprit (Barnosky et al. 2011). As we lose species through the loss of habitat, over-consumption, and persecution, populations become fragmented. Living in isolated remnants of their former range, such populations are unable to find new mating partners and territories in which to live, rendering them susceptible to inbreeding and loss of genetic diversity. In a situation analogous to the restricted matings that created the different breeds of dog, these isolated small populations can find themselves with such limited genetic diversity that they cannot respond to emergent diseases. In addition, they find themselves predisposed to genetic disease. One of the better-known examples of this phenomenon is that of the Florida panther, an isolated population restricted to a small region of southern Florida. As the number of individuals in the population declined markedly, an increase in congenital abnormalities was noted, including kinked tails, a whorl of fur on their back, and reduced reproductive fitness (Johnson et al. 2010, 1642). Genetic studies with cougars in nearby states revealed that the Florida panthers were closely related, allowing wildlife managers to introduce eight females from Texas to restore the panthers' genetic diversity. The reintroduction was a success, with reintroduced animals breeding with the remnant population, the population size increasing, and the disappearance of those physical abnormalities in the new offspring (ibid., 1644). This story is similar to that of the Dalmatian and pointer. In each case, there were reasonably close relatives that could be bred into the population to restore genetic fitness.

But what happens when there is no other population to draw upon? What happens when conservationists are trying to save the last few members of a species that are threatened by a genetic defect, or by lack of variation in genes governing immune response? Gene editing potentially offers a way to enable certain endangered species to overcome these challenges. However, here the situation is more complex, both ecologically and ethically, than in the case of domestic dogs. Endangered species are presumably going to live in the wild. A change in their genetic makeup could thus potentially have impacts on the ecosystem. This risk of negative impact must therefore be evaluated against the likely risk to the entire ecosystem if that species is lost.

For these scenarios, the distinction between gene editing and repair becomes relevant again. If the introduced changes in the genome restore functions lost because of human activity, then my default position would be that the change is ethical. If the alteration required to save that species involved gaining a function from another species, then I would argue for a very careful assessment of the potential risks involved, both for making the change, and for *not* introducing that new genetic function. The first rule of medicine is "do no harm," so risk to the system must always be weighed more heavily than risk to the species. Such risks must be evaluated carefully; one must consider, on the one hand, what would happen if the introduced gene causes a change in the life history of the species (or its ability to act as a reservoir of disease, et cetera), and compare that to the impact of the loss of that species on the system, on the other hand.

The Ethical Pathway Forward: Where to Draw the Line?

Dogs would not exist if not for the modification over time of the wolf genome through the process of domestication. My chapter has argued that gene repair is only different from selected breeding in the precision of the genetic change and in the greatly reduced number of individuals and generations needed to accomplish that change. For these reasons, one cannot logically differentiate genetic modification via CRISPR-Cas9 from selected breeding, except if the modified or introduced gene were not to be found in any other dog. A distinction between gene repair and gene editing can thus serve as an important starting place in ethical discussions about genetic modification of dogs and, relatedly, in other species.

References

Acland, Gregory et al. 2001. "Gene Therapy Restores Vision in a Canine Model of Childhood Blindness." *Nature Genetics* 28: 92.

American Kennel Club. n.d. "Dog Breeds." Available at: www.akc.org/dog-breeds

American Society for the Prevention of Cruelty to Animals. 2016. "Pet Statistics." Available at: www.aspca.org/animal-homelessness/shelter-intake-and-surrender/pet-statistics

American Veterinary Medical Association. n.d.a. "One Health—It's All Connected." Available at: https://www.avma.org/KB/Resources/Reference/Pages/One-Health.aspx?utm_source=prettyurl&utm_medium=print&utm_campaign=mktg&utm_term=onehealth

———. n.d.b. "Veterinarian's Oath." Available at: https://www.avma.org/KB/Policies/Pages/veterinarians-oath.aspx

Bainbridge, James et al. 2008. "Effect of Gene Therapy on Visual Function in Leber's Congenital Amaurosis." *New England Journal of Medicine* 358: 2231–2239.

———. 2015. "Long-Term Effect of Gene Therapy on Leber's Congenital Amaurosis." *New England Journal of Medicine* 372: 1887–1897.

Bannasch, Danika et al. 2008. "Mutations in the SLC2A9 Gene Cause Hyperuricosuria and Hyperuricemia in the Dog." *PLoS Genetics* 4 (11): e1000246.

Barnosky, Anthony et al. 2011. "Has the Earth's Sixth Mass Extinction Already Arrived?" *Nature* 471: 51–57.

Bernardi, Gretchen. 2010. "It's All About the Spots." *Canine Chronicle.* Available at: http://caninechronicle.com/current-articles/editorial/its-all-about-the-spots

Bomassi, Eric et al. 2011. "Patent Ductus Arteriosus in a Family of Chihuahuas." *Journal of Small Animal Practice* 52: 213–219.

Chastant-Maillard, Sylvie et al. 2010. "Embryo Biotechnology in the Dog: A Review." *Reproduction Fertility and Development* 22: 1049–1056.

Concannon, Patrick. 2011. "Reproductive Cycles of the Domestic Bitch." *Animal Reproduction Science* 124: 200–210.

Denizet-Lewis, Benoit. 2011. "Can the Bulldog Be Saved?" *The New York Times Magazine.* November 11. Available at: www.nytimes.com/2011/11/27/magazine/can-the-bulldog-be-saved.html?_r=0

Frantz, Laurent et al. 2016. "Genomic and Archaeological Evidence Suggest a Dual Origin of Domestic Dogs." *Science* 352: 1228–1231.

Gavazza, Alessandra et al. 2012. "Estimated Prevalence of Canine Type 2 Von Willebrand Disease in the Deutsch-Drahthaar (German Wirehaired Pointer) in Europe." *Research in Veterinary Science* 93: 1462–1466.

Georges, George et al. 1998. "Animal Models." In *Blood Stem Cell Transplantation 1.* Edited by Josy Reiffers et al., 1–17. London: Martin Dunitz Publishers.

Handyside, Alan et al. 1990. "Pregnancies from Biopsied Human Preimplantation Embryos Sexed by Y-specific DNA Amplification." *Nature* 344: 768–770.

Hart, Lynette. 1995. "Dogs as Human Companions: A Review of the Relationship." In *The Domestic Dog: Its Evolution, Behaviour, and Interactions with People.* Edited by James Serpell, 161–178. Cambridge: Cambridge University Press.

Horowitz, Alexandra, et al. 2013. "The Ethics of Raising Purebred Dogs." *The New York Times.* February 12. Available at: www.nytimes.com/roomfordebate/2013/02/12/the-ethics-of-raising-purebred-dogs

Jinek, Martin et al. 2012. "A Programmable Dual-RNA-gGuided DNA Endonuclease in Adaptive Bacterial Immunity." *Science* 337: 816–821.

Johnson, Warren et al. 2010. "Genetic Restoration of the Florida Panther." *Science* 329: 1641–1645.

Kim, Min Jung et al. 2011. "Generation of Transgenic Dogs that Conditionally Express Green Fluorescent Protein." *Genesis* 49: 472–478.

Kukekova, Anna et al. 2012. "Genetics of Behavior in the Silver Fox." *Mammalian Genome* 23: 164–177.

Larson, Greger et al. 2012. "Rethinking Dog Domestication by Integrating Genetics, Archeology, and Biogeography." *Proceedings of the National Academy of Sciences* 109: 8878–8883.

Leonard, Jennifer et al. 2002. "Ancient DNA Evidence for Old World Origin of New World Dogs." *Science* 298: 1613–1616.

Lindblad-Toh, Kerstin et al. 2005. "Genome Sequence, Comparative Analysis and Haplotype Structure of the Domestic Dog." *Nature* 438: 803–819.

Lupu, Marilena and Rainer Storb. 2007. "Five Decades of Progress in Hematopoietic Cell Transplantation Based on the Preclinical Canine Model." *Veterinary and Comparative Oncology* 5: 14–30.

Marsden, C.D. et al. 2016. "Bottlenecks and Selective Sweeps During Domestication Have Increased Deleterious Genetic Variation in Dogs." *Proceedings of the National Academy of Sciences* 113: 152–157.

McPherron, Alexandra and Se-Jin Lee. 1997. "Double Muscling in Cattle due to Mutations in the Myostatin Gene." *Proceedings of the National Academy of Sciences* 94: 12457–12461.

Meurs, Kathryn et al. 2012. "A Splice Site Mutation in a Gene Encoding for PDK4, A Mitochondrial Protein, Is Associated with the Development of Dilated Cardiomyopathy in the Doberman Pinscher." *Human Genetics* 131: 1319–1325.

Nagashima, Jennifer et al. 2015. "Live Births from Domestic Dog (Canis familiaris) Embryos Produced by In Vitro Fertilization." *PLoS One* 10: e0143930.

Nicholas, Frank. n.d. *OMIA—Online Mendelian Inheritance in Animals.* University of Sydney. Available at: http://omia.angis.org.au/home

Parekh-Olmedo, Hetal et al. 2005. "Gene Therapy Progress and Prospects: Targeted Gene Repair." *Gene Therapy* 12: 639–646.

Reynaud, Karine et al. 2005. "In Vivo Meiotic Resumption, Fertilization and Early Embryonic Development in the Bitch." *Reproduction* 130: 193–201.

Schaible, Robert. 1981. *A Dalmatian Study: The Correction of Health Problems.* Available at: www.dalmatianheritage.com/about/schaible_research.htm

Schoenebeck, Jeffrey and Elaine Ostrander. 2014. "Insights into Morphology and Disease from the Dog Genome Project." *Annual Review of Cell and Developmental Biology* 30: 535–560.

Scott, Danny et al. 2001. *Muller and Kirk's Small Animal Dermatology, Sixth Edition.* Philadelphia, PA: W.B. Saunders Company.

Sermon, Karen et al. 2004. "Preimplantation Genetic Diagnosis." *The Lancet* 363: 1633–1641.

Singh, P. et al. 2015. "A Mouse Geneticist's Practical Guide to CRISPR Applications." *Genetics* 199: 1–15.

Thalmann, Olaf et al. 2013. "Complete Mitochondrial Genomes of Ancient Canids Suggest a European Origin of Domestic Dogs." *Science* 342: 871–874.

Tonomura, Noriko et al. 2015. "Genome-Wide Association Study Identifies Shared Risk Loci Common to Two Malignancies in Golden Retrievers." *PLoS Genetics* 11: e1004922.

Trut, Lyudmila. 1999. "Early Canid Domestication: The Farm-Fox Experiment." *American Scientist* 87: 160–169.

Wiedenheft, Blake et al. 2012. "RNA-Guided Genetic Silencing Systems in Bacteria and Archaea." *Nature* 482: 331–338.

Zou, Qingjian et al. 2015. "Generation of Gene-Target Dogs Using CRISPR/Cas9 System." *Journal of Molecular Cell Biology* 7: 580–583.

Chapter 9

Digital Enchantment: Life and the Future of Gene Editing

Gaymon Bennett

I try not to be too much of a materialist, or a life-ist. What's so special about life, anyway? . . . in many ways, biological evolution is inferior to the cultural kind, and [to the] technological kind . . . A virus is an interesting information string, but is it more valuable than an idea? An idea is a different form of evolutionary replicator; it's a different form of information. But why is the virus more or less valuable just because it happens to be encoded in DNA, rather than in neurons, or magnetic strips, or ink on paper? There are so many different potential forms of information, and what I'm interested in, literally, is informational patterns and complexity.

Kevin Esvelt, quoted in Chapter 3, page 62, this volume

My purpose in this chapter is to provide the reader of this volume with the cultural background and with the historical context for understanding the project of biotechnology today. This background and context deeply inform current developments in gene editing, thereby playing a defining role in the stated aspirations by gene editing scientists, as well as in the broader ramifications of their work.

Following other contributors in this volume, I hold the position that biotechnology governs how we live our lives. As surely as other major cultural institutions—not least law and politics—biotechnology inflects our sense of how life is constituted and how it ought to be ordered. I add to this position a further proposition: that the power of biotechnology cannot be successfully regulated if dealt with only in a piecemeal fashion. Biotechnology needs to be governed at the level of the broad cultural imaginaries that govern it. Hence the purpose of this chapter.

The term "cultural imaginary" (or "social imaginary," as it is more often put in scholarly literature) is a neologism that I will use throughout this chapter. This neologism was coined by political and legal theorists to bring into analytic focus the everyday practices and assumptions that must be in place for a group of people to make sense of the world together. One prominent philosopher puts it like this:

This approach [i.e., the study of social imaginaries] is not the same as one that might focus on the ideas as against the institutions of modernity. The social imaginary is not a set of ideas; rather it is what enables, through making sense of, the practices of a society.

(Taylor 2002)

In legal and political theory, the term "social imaginary" has been used most often to specify and make sense of the cultural formation of modern nation states, and the ways of engaging reality that allow modern societies to function at even the most basic levels. In this chapter, I want to use the term to make sense of the everyday background assumptions required for biotechnologists to go about their daily lives—from the ways in which they tacitly imagine the nature of living things in the design of research programs, through how they order their experiments and equipment in their labs, to how they envision, promise, and attempt to deliver a future in which engineered living things make life better.

I propose that the cultural imaginaries of biotechnology—or "the biotechnical imaginary," as I will sometimes call it in this chapter—has a direct bearing on the ways in which biotechnology today "empower[s] some forms of life . . . making them natural while others, by comparison, come to seem deficient or unnatural" (Jasanoff, Hurlbut, and Saha 2015, 26–27). The biotechnical imaginary functions as a reservoir that is drawn from "when a society takes stock of alternative imaginable futures and decides which ones are worth pursuing and which ones should be regulated, or even prevented" (ibid.). Biotechnical imaginaries, in short, shape the everyday practices of biotechnicians, enabling them to make sense of their world and their work. But they also shape our collective future by functioning as the background reserve that comes into play in moments of collective decision making.

Over the last decade, among those empowered to envision, plan, and facilitate biotechnology, no single set of actors has arguably had a greater influence on our collective sense of which "alternative imaginable futures . . . are worth pursuing" than a group of biotechnologists who call themselves "synthetic biologists." Among self-identified synthetic biologists, it is also fair to argue that the most influential subset has been a group of individuals and institutions that can be loosely called "digital biologists." In what follows, I will explain what synthetic biology is, and provide a close examination of the biotechnical imaginary that informs digital biology. Let me note here that whereas synthetic biology has looked to computer engineering as a general model for biotechnology, digital biology has intensified those efforts by explicitly embracing the idea that living things can be treated like digital code. The reason for this examination—and hence the goal of this chapter within this volume—is to help the reader become clear about the ways in which the biotechnical imaginary is shaping current developments in gene editing, and CRISPR and gene drives in particular.

One last word of introduction. In what follows, I will argue that one of the defining features of the modern biotechnical imaginary is that it perceives the living world as "disenchanted"—that is, fundamentally passive and mechanical. I will explain in more detail below what I mean by "disenchanted," and why this concept is crucial for understanding gene editing. I flag it here, because I think the commitment to disenchantment in biotechnology is becoming increasingly difficult to sustain. Which is another way of saying that biotechnology today, or at least those aspects of biotechnology that I call "digital biology," is raising the question of the "re-enchantment" of the living world, and thereby potentially unsettling the dominant biotechnical imaginary. This matters deeply to the regulation of life, in both the legal and biological senses. It is the question of re-enchantment and the regulation of life that is fundamentally at stake in gene editing, as I hope this chapter will make clear.

Biology and the Digital Imagination

In early 2016, news broke that a handful of biotechnologists had quietly begun assembling the talent, technologies, and resources needed to manufacture a fully synthetic human genome. And though the biotechnologists expected blowback (they held their planning meeting in secret), leaked news of their plans produced a relatively muted response.

The absence of outcry was surprising. The specter of engineered modifications to the human genome has haunted the ethics of biotechnology for half a century. Perhaps the response was due to over-saturation. After all, these same technologists have been promising for 20 years to routinize the rational design and construction of new living things. News that they were now ginning up a "Human Genome Project 2.0" (as the biotechnologists proposed calling it) could be read as just one more installment in a familiar story.

The organizers of "HGP 2.0" included some of the most mediagenic figures in an area of biotechnology known as synthetic biology. This group also forms the core of what is now being called "digital biology." For the purposes of this volume, digital biology can be thought of as gene editing taken up on the level of the whole genome, i.e., editing involving the whole complement of DNA in an organism. What makes it "digital" is that it names a mode of synthetic biology that holds that the long-standing scientific task of understanding how life works can now be productively sublimated to the technical task of building computers powerful enough to make life work differently.

Because self-identified synthetic biologists-cum-digital biologists play such a significant role in reshaping the biotechnical imaginary today, it's worth taking stock of the sensibilities and aspirations underwriting the planning of HGP 2.0, regardless of where the project ultimately goes. Above all, it is important to note the manner and extent to which digital biologists have come to believe that biology is the material basis of a new high-tech revolution, and that this revolution is being (and only can be) realized through an ongoing synthesis of living and computational things. The epigraph from Kevin Esvelt above is exemplary in this regard.

This synthesis of the biological and computational is, on one level, straightforwardly material: without computers to catalogue, sequence, store, parse, rework, and recapitulate data, biotechnology today simply doesn't work. But if it is material much of the time, this synthesis is a fixture of the biotechnical imaginary all of the time: the presumption of deep connections between the biological and computational enable biotechnicians to make sense of their everyday practices. Indeed, for insiders, the future of biotechnology as a computer-mediated art of making new living things ("so many different potential forms of information") has become so self-evident that doubts about its inevitability are unthinkable.

This attitude of self-evidence is underwritten by the presumption that the logic of living things and the logic of digital things are fundamentally commensurate. Consequently, many technologists (some inside, some outside biotech) take it as a matter of course that the masters of the digital world, who brought you the personal computer, machine learning, big data, and self-driving cars, can and will become the masters of the biological world. A theory of life gets linked to a theory of power.

Broadly speaking, the biotechnologists working to pull together an HGP 2.0 have focused on two primary technical goals, both crucial to a vision for integrating the biological and the computational. The first concerns software programs for the rational design of novel organisms; the second, the construction of machines for their automated assembly. For example, George Church, a prominent figure in gene editing and a convener of HGP 2.0, has long promoted the idea that biotech, like high-tech, needs dedicated "fabrication facilities," i.e., facilities for the automated manufacturing of custom designs. Such facilities, he and others have argued, would allow technicians to focus on higher-level design problems, allowing machines to take over the tedious labor of making working designs work. Approaching from another side, software engineers, like those at the company Autodesk (who also helped convene HGP 2.0), have focused their efforts on building computer-aided design software for biology. Autodesk assumes that designers alongside scientists will eventually drive biology.

This approach to biology as a problem of automation and design is obliquely captured in the subtitle of HGP 2.0: "HGP-Write." The "write" in HGP-Write is both a provocation and a possibility. It suggests that the original human genome project "merely read" human DNA, without yet equipping technicians to "write" into existence new kinds of living things, which, for digital biologists, is the whole point. The "write," however, is also an expression of the self-confidence of digital biologists. This self-confidence expresses itself in the belief that the core problems of biology—from understanding complexity to engineering novel organisms—can be given over to the arts of computer engineering. Recall Esvelt's statement in the epigraph that "there are so many different potential forms of information and what I'm interested in, literally, is informational patterns and complexity."

The phrase "digital biology," it bears noting, predates these current expectations. It first began to be used in the 2000s to refer to the massive influx of computer-mediate data into the daily life of biology, an influx that followed in the wake of the genome sequencing projects of the 1990s. In the case of HGP-Write, however, "digital biology" means something more precise. The vision for HGP-Write is predicated on the eventual ability to design, *in silico*, a whole human genome, print it out using synthetic (i.e., non-living) chemistry, and assemble it in a cell that has had its own DNA removed.

Although technicians are years from being able to do this in human cells, the J. Craig Venter Institute (JCVI) spent the first decade of the twenty-first century establishing the reality proof for this new sense of digital biology. In 2010, JCVI announced that its scientists had successfully produced in bacteria what they called "synthetic life." JCVI scientists had spent a decade designing a "minimal genome"—a bacterial genome that had had all DNA which did not seem essential to cellular survival excised—and had successfully "booted up" that genome *in vivo*. This success was widely trumpeted as digital biology: after all, the resulting bacteria did not exist until it was designed on a computer and printed out.

Despite all JCVI's innovations, and despite the gene editing ambitions of HGP-Write, this new sense of digital biology overlaps with previous uses of the term in one fundamental respect. From the genome sequencing projects of the 1990s forward, biotechnicians have undertaken their daily work on the assumption that the logic of life is commensurate with the logic of digital computation. This assumption is not philosophical. It is, rather, part of the operative rationality of laboratory life: living things are approached by way of sequence information, made available through online databases, and recapitulated in labs around the world through synthesis. Living code is being managed as digital code—"bits to bytes to bits," as one bioengineer has put it.

Yet there's a twist: for all the talk of life as information, experimental work in the lab actually privileges a different mode, one calibrated less to an *a priori* digital concept of living things and more to an experience of their elusive and wily properties. The labor of constructing purpose-built organisms may rely in fundamental respects on digital data and computer-run processes, but it remains largely experimental, which is to say reiterative and unpredictable. Such reiterative design and construction is certainly systematic and regimented, even highly controlled. But it is highly controlled precisely because it is clear to experienced laboratory technicians that there is only so much one can do to anticipate the form and function of new living things. The computational and the biology may be getting synthesized, but on experimental grounds the biological continues to frustrate the computational.

This matters because, although there is a disjunction today between a digital conception of life and the laboratory experience of that conception, that disjunction usually is cast as one of relative immaturity. The idea is that experimental conditions have simply not yet caught up with high-tech ambitions. Talk of immaturity, in other words, trends toward the idea that biological design should continue to be carried forward in the mode of digital design, but that we are simply not there yet. A closer look at the everyday tensions of laboratory life suggest that something more is going on. Rather than experiential immaturity, the disjunction signals that talk of computers and living things is much more limited than biotechnologists might suggest and thus may only take biotechnology so far.

From Digital Biology to Gene Editing

Describing his scientific sensibilities to legal ethnographer Irus Braverman, Kevin Esvelt put things this way: "I try not to be too much of a materialist, or a life-ist. What's so special about life, anyway? A virus is an interesting information string, but is it more valuable than an idea? An idea is a different form of evolutionary replicator; it's a different form of information" (quoted in Chapter 3, page 62, this volume). On one level, Esvelt's formulation is not particularly unusual. After all, he gives voice to an opinion shared by many technicians, and one that is tacitly widespread among digital natives. The opinion shouldn't be mistaken for a devaluation of life per se. Rather, it might be thought of as an elevation of other forms of experiential existence, as Braverman points out in her chapter. Nonetheless, the formulation is striking. And it is striking not only because it unsettles the value of living things, but because it stands in tension with the lived experience of practicing biology, both within gene editing and beyond (see Chapter 3, this volume).

In his writings on the concept of life in biology, the historian Georges Canguilhem offers a provocation, namely that the ghost of Aristotle will continue to haunt the life of science and the sciences of life long after scientists and technicians have stopped paying attention (Canguilhem 2000). The reason is that Aristotle was the first philosopher of biology to grasp that life is a syllogism. According to Aristotle, the concept of the living thing—its essential nature—is the living thing itself. That is to say, the living thing orders itself according to its own internal principles. This made Aristotle an empiricist: if life is its own concept, then if you want to know what a living thing is, you actually have to go look.

To say that the concept of the living thing is the living thing itself is to say something less obtuse than it might sound. Concept, we recall, derives from the same root as conceive. It means "to be born" as well as "to grasp." When a living thing is "born" into the world, it grasps itself in itself as itself. Life, Aristotle argues, is defined by an immanent activity—nature defines its own nature. Or, as Spinoza subsequently put it: nature natures itself. Life is self-moving, and thus has a self-referential character (Spinoza 2002).

Aristotle taught that this immanent activity of a living thing—what he referred to as its "soul"—can be discerned in its form. Aristotle believed this immanent activity showed itself in the way a living thing maintains its form (allostasis) against an onslaught of external forces (Aristotle 1984). Aristotle thought it was yet again more apparent in the way a given form, a soul, persists between generations. The concept of a thing is passed on through conception. The catch, which Canguilhem brought into sharp focus, is that the mechanisms of allostasis sometimes fail and heredity produces abnormalities. The soul, in these cases, does not reproduce itself—it does not conceive itself as itself. Life is not always "grasped" by its concept. And yet in failing to grasp, life may change—mutate or evolve—in a durable, that is to say survivable, manner. Life, Canguilhem underscores, is capable of error.

What this means for Canguilhem is that it is never enough for the sciences of life to be based on a concept of life. The sciences of life must also be based on an experience of life. The combination of these two—the concept and the experience—generates a situation in which biology can never quite free itself from what has been called "vitalism." Vitalism, here, does not mean (as it is sometimes assumed by critics of vitalism) that life is a combination of a material body and an immaterial soul. The soul, rather, is the form of the body. Vitalism thus refers to the way in which the living thing functions as a unified being that has an intrinsic capacity to act in the world in a way that is neither predetermined nor random. Living things are not just bundles of passive mechanisms; they are also agents.

It is for this reason that despite several centuries of mechanical metaphors in biology (including metaphors of life-as-information), it remains the case that when biologists study how living things exist in the world they do this by taking the fact that living things are alive as a conceptual *a priori*. The seemingly banal idea that living things are alive is actually the first precept that allows for the logically subsequent manipulation and analysis of life.

The tensions and interplay of "life as a concept" and "life as an experience" can be seen in the conflicting approaches to engineering that have riddled "gene editing" over the last few years. Take, for example, work on the so-called CRISPR-Cas9 system. As has been widely discussed, in late 2012 the biochemist Jennifer Doudna and her research team in Berkeley, along with Emmanuelle Charpentier and her research team in Sweden, published breakthrough work on the CRISPR system. This simple but powerful technology harnesses the inimitable capacity of a living thing to manage the potentially hostile conditions of its milieu. It does this by capturing genetic material from invading viruses and passing it on to its immune system for future use. CRISPR is one function of an agent-like being.

In its reengineered form, as Doudna and Charpentier showed, the system can also be used to make highly precise cuts and specified insertions in DNA, including double-stranded DNA. This engineered capacity has quickly made CRISPR a foundational tool for building designed DNA sequences in the cells of a wide range of organisms, including—as Feng Zhang at MIT and George Church at Harvard first demonstrated—human cells. The Doudna group demonstrated that an elegant technology renders the painstaking work of editing genes (relatively) more accurate, reliable, fast, and cheap. The Church group, meanwhile, demonstrated that this technology is portentous with regard to the future of human self-elaboration. Together, these two groups produced the widespread sense that CRISPR was the technology needed to carry life across the threshold of imagined biotechnical futures.

The story of CRISPR, in its early moments, was a story of biotechnology keyed to the experience of living things. The power of the system was precisely that it allowed for a relatively simple intervention in otherwise complex and dynamic systems. Using CRISPR one did not need to control the entire organism, as sometimes seems to be on offer in digital biology; one only needed to inflect it. This inflection is powerful precisely because it allows the engineer to leverage the potencies of living things—i.e., their ability to behave in ways that are neither predetermined nor random (Riskin 2016). Or, to put it negatively, this inflection is powerful because it does not treat living things as passive. It rides on the back of their ability to sense their environments, discriminate, move, communicate, relate, adapt, coordinate, and choose.

On the surface of things, then, the CRISPR approach to the experience of life stands in stark contrast to the concept of life that infuses the project of digital biology. Digital biology proceeds on the assumption that life can be made to operate as a set of passive mechanisms. The goal is to transform these operations wholesale through regimes of increasing technical complexity (hence the call to invest a decade of funding in the making of an HGP 2.0 through design and automation). Gene editing via CRISPR, by contrast, approaches living things as self-regulating complexities, which can best be transformed through the insertion of simple and clean, but therein potent, technologies.

These potential tensions between CRISPR and digital biology have been resolved in a political manner. Where CRISPR began as a story of technical capacity keyed to the experience of living things, it was quickly remade into a story of the reification of the regnant conception of life underwriting digital biology. The CRISPR breakthroughs were almost immediately taken up by the digital biology publicity machinery, and made to reinforce the already-diffuse sense that biological engineering has become a digital designer's playground.

CRISPR, that is to say, was mobilized under the banner of the hybridization of bio and tech. And despite apparent differences in design sensibility, the mobilization is not particularly surprising. George Church, after all, was one of the first biotechnicians of high standing to endorse the digital biology brand. And though Doudna did not obviously have anything to do with that brand prior to her CRISPR breakthrough, she quickly began working with leading figures in digital biology, publishing papers and founding companies.

An otherwise clear difference in sensibilities was sublimated by a shared ethos: namely, the sense that digital possibilities have become the *sine qua non* of biological futures. "Writing" and "editing," after all, are both metaphors indexed to code and not text. And as metaphors indexed to code, they both call into being a hybrid figure of biotechnology, one where the artisans and industrialists that brought us the iPhone and the Cloud will provide the institutional knowhow needed to turn life into the next tech revolution.

Synthetic Biology and the Figure of Information

The regnant ethos that shapes current developments in contemporary biotechnology is not simply a function of a "hype" machinery, which allows technology to operate in a space of cultural cachet. It is, rather, an orientation toward the future, which both establishes possibilities and governs accountability. In the case of digital biology, this orientation is largely a function of its deep connections to the broader and slightly older movement in biotech called synthetic biology. Synthetic biology can be thought of as encompassing both gene editing and digital biology. The governing ethos in synthetic biology turns almost entirely on the elusive and seductive figure of "information." "Information" has, of course, been a commonplace of science and engineering for more than half a century. It has remained a commonplace, however, because though conceptually under-determined, it continues to unleash pragmatic energies.

The term has old roots. In medieval Christian thought, it was a term of inspiration, in the literal sense of "breathing in." To be informed was to be filled by the spirit—to be breathed into, inspired, animated. In this sense, the term was used to name the ways in which ensoulment holds together vital elements of a body (the blood, the heart, the limbs, the brain, and so on) and makes them a single living organism. By the late-nineteenth century, the concept, if not the word, had made its way into biology. It came into force through conversations about heritability. One then-dominant question was how a living being, for all its seemingly mechanical parts, could maintain coherence in structure and function from conception, through the course of its development amidst the pressures of allostasis, and finally, pass that coherence on to its progeny.

A handful of nineteenth-century biologists debated the possibility of a "primitive impulse, a primitive action and message, which nature repeats according to a pattern determined in advance" (Canguilhem 2009, 77). It would be a stretch to say that these biologists were seeking to reproduce an Aristotelian notion of soul by another means, but it is fair to say that they were bent on discovering a biological reality that could provide a physiological equivalent of the soul. Such an equivalent, it was thought, would allow biologists to bind together a sequence of biological states from birth, construction, growth, restoration, or reproduction and conceive of them as a single reality without appeal to either strictly mechanical or vitalistic metaphors. That is to say, it would allow for something like vitalism—the immanent activity of life—without resorting to a metaphysical dualism wherein life was a combination of a material body and an immaterial soul.

If information operated in nineteenth-century biology as a hypothetical reality that allowed biologists to conceive of living things in their physiological continuity, in the twentieth century it was famously reasserted as a material reality that allows for the conception of living things as forms of mathematical and linguistic possibility. With the discovery of the structure of DNA, and the proposition that the key to both heritability and development lay in the ordering of a finite number of chemical bases, life was reimagined on the metaphors of information theory—programs, codes, and instructions.

Taken as a material reality, information was neither static nor *sui generis*. And the question of whether it was the form of an immanent activity or the articulated structure of a passive mechanism awaiting activation by outside forces, remained unanswered. Information clearly was activated by interactions with an environment—above the entire cellular environment and the environment of the cell. But those interactions were regulated by the code of the living thing itself in such a way that the organism could assert itself against its environment.

Metaphors of code thus allowed life to be imagined, studied, and encountered as a series of communicative operations wherein code is made manifest as non-living chemistry. It could be approached, that is to say, as an utterly synchronic reality, in which all the information needed to make and maintain an organism was present and ready for expression. Yet life could also be imagined, studied, and encountered as an agent capable of expressing itself, and expressing itself variously—that is, as living information. Life presented itself as a diachronic potential, whose past was as vital to its meaning as its present and future.

To say then that the early dreamers of synthetic biology were inspired by the elusive figure of information is to take note of several ontological factors that defined the initial direction of their work. Above all, they were inspired by the multiple ways living systems are able to give form to, and process signals pertinent to both their inner states and outward relations. This responsive variability could be seen as effervescent from the perspective of information theory: vibrant materials capable of an endless variety of states, functions, and interactivities.

The living world could thus be read as a kind of vital "basis set": a limited number of motifs that could, in principle, be endlessly transacted and transformed to produce an unlimited number of biological operations. Moreover, these potentially unlimited operations could be said to have as their predicate and point of reference not the living organism per se, but the materiality of the DNA—or at least the DNA as the material basis of an ensemble of informational bits that could, in principle, be taken up as discrete, predictable, and interoperable.

Like expressions of information produced through programming, this materiality seemed to have a potent combination of conventional and necessary properties. Conventional in that biological forms and function could always be otherwise. Necessary in that any given form or function seemed to have a stable and regular basis, which could be counted on to behave as programmed. It was ultimately this second property that caught the imagination of engineers—though it would not have been attractive without the first.

Synthetic Biology and the Thought of the Living

A decade and a half has passed since a small group of computer engineers and biochemists first took up the name "synthetic biology." In that time, the brand has boomed. It has drawn in a wide array of practitioners from disciplines and institutions around the world. Given its breadth and heterogeneity, there are any number of ways the story of synthetic biology could be told. Any plausible version of that story, however, must take account of developments in the Artificial Intelligence Lab in MIT and in the Molecular Sciences Institute in Berkeley in the early 2000s.

A principal figure on the MIT side was Tom Knight, a career researcher in MIT's AI Lab. As others have detailed, in 2002, Knight and a cohort of graduate students in bioengineering began to outline a program for rethinking molecular biology on the model of computer science (Balmer, Bulpin, and Molyneux-Hodgson 2016). Knight had, by that point, exited from a lifetime's work of designing computer systems. He had become committed to answering the question of how, and to what extent, living systems could be realized as computational systems. Knight's operative assumption (shared by others) was that cellular operations could be reimagined on the model of circuits, and the engineering of DNA on the arts of digital programming.

Knight arrived at MIT as an undergraduate in the late 1960s and stayed on as a research scientist. He came at a moment of re-intensified interest in the life of information and its possible relationship to the life of technology. Debates had begun to crescendo over the limits of cybernetics and whether and how advanced mathematics could continue to transform communications theory into computational power.

By the mid-1970s, two questions became important. The first concerned pedagogy. Which training regimes would best help young technicians become engineers capable of designing and making controlled informational systems—such as operating systems and circuits? The second concerned politics. Which forms of communication among and between makers might best facilitate the growth of a distributed community of technicians? Arguably, the most impactful answers to these two questions were formulated under the guidance of Hal Abelson and Gerry Sussman. From roughly the 1970s through the 1980s, Abelson and Sussman became the major architects of a growing design philosophy predicated on the idea that information systems *qua* communication systems ought to be designed in such a way that other designers, working in other places and at other times, would be able to understand and directly add to those systems.

This commitment to distributed engineering had the effect of making the pedagogical question coincide with the political question. Abelson and Sussman proposed that a computer language is not only a means of making a computer perform a set of technical operations. It is also a regularized medium for expressing methodological ideas, that is, ideas about how best to create relationships between different kinds of operators.

New expressions articulated using a computer language are, in this sense, as social as they are technical: everything expressed is both the outcome and predicate of other people expressing other things. The power of code could certainly be imagined synchronically as a set of functions available to anyone capable of understanding the code. But it could also be understood diachronically: as a developmental history of a recursive relation between information systems and their makers.

On this model, makers of information systems proceeded on the operative assumption that information is not materially bound, per se. In practice, this meant that the only limits on the things one could express through digital logics were limits imposed by the designer's own imagination. This sense of seemingly unlimited possibility within the space of a rule-bound system produced a distinctive maker's sensibility: the sense that when it comes to programming, genius and savoir-faire (both individual and collective) are the primary passage points to a transformed future. Such a sensibility put a premium on the virtue of collective action: the need to invent modes and forms of practice that would equip a community of users to work and rework one another's creations. This sensibility was primed for the early synthetic biologists.

The conspicuous wrinkle in this sense of unlimited potential were the physical constraints of a computer's hardware. It is not surprising that Knight and others at the AI

Lab remained intensively engaged in the task of rethinking and redesigning integrated circuits as the material bases for the immaterial expression of digital ideas.

By the mid-1980s, Sussman, Abelson, and Knight were actively thinking about what could be done for the future of digital processing when (not if) the computer industry reached the physical limits of processing power. Sussman and Knight began exploring whether it might be possible to reimagine living cells as possible computational platforms. Their back-of-the-envelope calculations suggested that the processing power of even a simple single organism exceeds the physical limitations of any chip they could imagine. The question was whether the analogy between information processing in computers and the information processing of living things could be made literal.

In the mid-1990s, Knight undertook to learn the biochemistry he thought he would need to explore this possibility. Along the way, he recruited a cohort of graduate students and postdocs to take up these efforts in a sustained fashion. Basic to Knight's program was the proposition that in order to transform biology into technology, technologists would need a feasible means of imagining and rendering living systems as series of standardized components. This proposition became definitive for the early synthetic biology manifestos, though it was never taken up in a sustained fashion by other major players. For Knight and his cohort, however, the idea of standardized parts was indispensable to their biotechnical imaginary. From their point of view, the creation of standardized components in computer engineering was the major structural transformation that allowed engineers to shift intellectual energies from the slow work of building the material bases for programming to the redesign and redeployment of those bases.

Less evident, but ultimately more important, this vision for synthetic biology inherited from computer sciences a feel for the limitless possibility of information. This helped generate a cultural situation in which the functions of living beings could be talked about and imagined without reference to the living beings themselves. The vision of cells-as-assembled-components entailed an ontology wherein the living being was assumed to be nothing but a series of contiguous juxtapositions across multiple interactive scales.

In all of this, the figure of information remained integral but opaque. It was integral in that the transformation of information into function was, on one level, the whole point of the enterprise. It was opaque because no one thought much about what this might mean biologically. It was taken for granted that information would persist in and across the interfaces and interactions of material components. An underexamined ontology of information proved significant. It helped engineers continue to imagine processing of information as a more-or-less unbounded enterprise. Informational expression might be limited by material bases (code dependent on hardware), but the transformation of information through encoded expressions could continue as more or less unlimited—except, of course, where imagination failed.

Despite the limited success of component-based engineering, synthetic biology continued for a decade to be imagined as an enterprise of unlimited instrumental payout. If the design of new functions in living systems was unlimited in principle, then there was also, in principle, no set of biological processes that could not be redeemed by way of this approach to the world. Technical criticisms of synthetic biological projects could be countered with a standing reserve of hope and expectation, a hope and expectation rooted in the ontology of information. It is hardly surprising in this light that synthetic biologists have been willing to take up the mantle of digital biology as a seemingly inevitable next step in their enterprise.

The living being within which the designer's operations are physically placed, remains, at best, a problem to be dealt with. The organism, after all, wants to spend cellular energies doing things that seem to have little bearing on the engineered functions. Attempts to "brew up" yeast, for example, which have been made to spend almost all of their energies producing a targeted molecule, are regularly frustrated by the yeast's efforts to make itself live, or, conversely, its tendency to die. The seemingly stable set of operations encoded in designed DNA, which are meant to serve as the instructions for targeted functions, is constantly at risk of being subverted by the wily transformations of the living being into which these operations are set into motion.

From the point of view of synthetic biology, however, there has never been a living being that might serve as the reference point for designed operations. Operations are simply hierarchically arranged motifs captured in encoded components. Yet this lack of an organismal referent has proven to be a sticking point at the heart of the enterprise. After all, from the outset, the goal of synthetic biology was to take advantage of the effervescent capacities of living beings—their ability to interact, respond, adapt. The vitality of the living being was the attraction. Life preceded technique. And yet it was precisely the vitality of the living being that has been conceptually externalized and thus has had to be continually contended with on experimental grounds.

Hence my suggestion above that Esvelt's sensibilities are not particularly unusual. To repeat the point, he not only gives voice to an opinion shared my many technicians, but also admits to sensibilities that are tacitly widespread among digital natives. Whether and how these sensibilities, anchored in a concept of life, can be squared with an experience of life remains to be seen.

Gene Editing and Life's Disenchantment

Over the past several decades, a body of work has begun to take shape in the social sciences, which is sometimes referred to as "the ontological turn." Underpinning this work is a shared commitment to critically reappraising the idea—long fundamental to the human sciences—that social constructions are the defining feature of human realities (Kohn 2015). As part of this critical reappraisal, "the ontological turn" rejects an implicit dualism that cleaves the human from the nonhuman, casting the first as vital, active, and world-making, and the second as mechanical, dull, and passive. It equally rejects the assumption that the construction of social realities is an exclusively human practice and not also an intrinsic capacity of nonhuman things—including the kinds of organisms that are being constructed as part of regimes of gene editing.

Scholars of the ontological turn (e.g., Eduardo Kohn, Anna Tsing, and Donna Haraway) have adopted a view of reality that can no longer be squared with what the classic political economist Max Weber referred to as the "disenchantment" of the modern world—or, translated more literally, the modern "de-magification" of reality (Weber 2015). A "magified" or "enchanted" reality can be thought of as one in which all of life is experienced as "immanent to itself," i.e., where living things participate in the co-constitution of one another's being—the human in the nonhuman, the living in the dead, the animate in the inanimate, and so on (Weber 1958). Disenchantment thus names the rending of such vital immanence, tearing reality into a raft of dualities. Conspicuous among these is the split between humans, understood as bearers of agency, and a nonhuman, nonagential world (Kohn 2015).

Scholars of the ontological turn reject disenchanted dualism as metaphysically over-committed. The reasons for this are complex and diverse. But it is fair to say that as part of a commitment to empirical inquiry, these scholars remain open to the possibility that nonhuman things may be as conventional as human things. As creatures of convention, these nonhuman things may exercise a good deal more force on forms of human living than moderns may have suspected. Humans may find that in trying to construct reality, including biotechnical reality, reality pushes back.

Consider the goal at the heart of digital biology of producing computer-aided design (CAD) programs for biotech. CAD programs treat biological materials as digital materials, and thus as discrete, separable, and passive. The ways in which those programs assemble, interface, categorize, characterize, and—the designer hopes—recapitulate living things, is simply presumed to be sufficient to the (re)making of living things. CAD programs, in other words, conceive of living things in digital terms, and take the being of the one to be enough like the being of the other that computer software will successfully function as a medium through which rational design and automated assembly become possible.

At the same time—and this is crucial—the artefact of such analogical reasoning pushes back. During that push back, digital conceptualizations fade to gray as the lab technician strives to adjust and readjust the living thing according to previous plans. In the midst of those responsive engagements, an "analogy of being" begins to run the other way: life is not only understood in terms of digital logics; digital logics must also be reimagined under the pressure of the logic of life, a logic that makes itself known through a refusal of cooperation. The ongoing need for experimentation in biotechnology—despite dreams that life will eventually allow itself to be simply edited and printed using design-driven technologies—gives testimony to what Charles Peirce called the "outward clash" between humans and that which lies beyond their forms of representation (2009).

This clash is one possible outcome of the aporia—a seemingly irresolvable internal tension—that marks both living things and the sciences of living things. It is the aporia between life as a concept and life as an experience—the difference between life in synchronic and diachronic modes. The crucial variable that allows this aporia, these two modes, to avoid being merely a lacuna or blockage is what medieval philosophers used to called an "analogy of being." This is the idea that one thing is so much like another that it participates in and thus partially constitutes the existence of that other thing—the two things share a "participatory ontology." In the case of biotechnology, the analogy of being concerns digital things and living things: they participate in one another's existence because they are both informational.

It is vital to keep in mind, however, that an analogy of being works in this case only insofar as it helps sustains conceptual and experimental motion *between* life as a concept and life as an experience. The analogy needs to remain supple and open. When it becomes fixed, analogies give way to identities and the motion freezes up. And this, I submit, is precisely what is happening in biotechnology today.

For more than half a century, talk of life and computers, or life and information, remained slack, allowing experiments to continue even where a biological understanding of molecular systems has been lacking. Along the way, however, the analogies became so deeply internalized into the rhythms of laboratory life that infrastructures—from buildings to habits, processes, concepts, machines, and business plans—were designed to actualize and thus reify them. Slowly but surely, talk of similarity has progressively given way to the presumption of identity. Despite obvious morphological and material differences an inner identity between the nature of living things and digital things has come to be operationally taken for granted.

Yet the presumption that living things are like digital things, and therefore the presumption that they can be treated as discrete and passive, is only that: it is a presumption. In practice, the technicians tasked with constructing purpose-built organisms are constantly having to renew this presumption in order to keep things going. This takes the form of renewing a commitment to disenchantment, and whether or not that commitment remains worth the energy it costs in terms of experimental adjustment is far from clear.

Digital Biology After Disenchantment

It is curious that digital biology has been claimed, at least by the framers of HGP 2.0, as a bookend on the genome projects as well as a threshold beyond them. It is curious because of what the genome projects have done to the daily life of biotechnology. The projects reasserted, within a space of biotechnical reason, the idea that life not only has an experiential history; it *is* an experiential history.

Typically, genomics is defined as the study of the full complement of DNA in a given organism. Yet it can also be defined as a computer-enabled manifestation of the genetic residues of the evolutionary history of living things: an art of making visible the manner and extent to which living things are a material expression of living histories. Genomics, in other words, is the scientific articulation of an old human practice: it reveals yet again how the living are vital expressions of the dead—i.e., a broad web of prior living things. It thereby opens the possibility of using those vital expressions to change the course of the living. Genomics gives expression to the participatory nature of living things and makes that participatory nature available to biotechnology, including gene editing.

Digital biology can be seen in this light as both a contraction and an exit. It is a contraction because, while it recognizes that genomic information is an accretion of lived histories, it takes up those lived histories as if they can recapitulated in the form of digital annotations, characterizations, and quantifications—a usable synchronic remainder. It is an exit, because that synchronic remainder, it is believed, can be expressed as the encoded basis for the composition of something that will begin a new history—"different potential forms of information," to quote Esvelt again (Chapter 3, page 62, this volume).

In all of this, what has not yet been taken seriously as a design parameter, let alone a material reality, is the extent to which biotechnology is quietly trying to manage the tools and knowhow needed to conceive of life as enchanted—that is to say, life as a reality which is immanent to itself in the Aristotelian sense outlined above. Evolutionary and environmental scientists incorporated the participatory ontology of life decades ago. But these participations are only now making themselves visible on the biotechnical scene. The question of whether they will continue to be an externality and source of experimental frustration or whether they can be converted into life beyond the digital remains unclear.

Talk of information in biology is nothing new. The institutional consequences of this talk, however, remain unsettled—not least because notions of information at play in high-tech are likewise unsettled. Nonetheless, talk of biology and information tends to take for granted a working concept of life that is no longer self-referential, that is to say, one where life is no longer defined on its own terms. Instead, the computational stands in for, and thereby displaces, the biological.

The shifting notions of information at play in biology are freighted. Most important for this chapter, these notions teach biotechnologists to treat living things as passive things—taking "passive" in the sense offered by Jessica Riskin, namely as lacking an intrinsic capacity to act in the world in a way that is neither predetermined nor random (Riskin 2016, 6). Because information is taken to be passive—to be rule-bound, mechanical, and determined—biology becomes so as well. This presumed passivity has become a tacit warrant for biotech's confident claims that it will (continue to) deliver on what Max Weber once referred to as "this-worldly salvation"—the belief that science will inevitably deliver health, wealth, and happiness (Weber 2015). The idea that biology is fundamentally passive seems to assure that once biologists figure out how living things work (or, rather, how they can be made to work), engineers will be able to carry out any operation consistent with the material and mechanical conditions of life.

The problem is this: if the presumption of passivity is biotechnology's warrant, it may also be its limit. The salvational potency of biotechnology turns entirely on biotech's ability to leverage the biological potency of living things. But the potency of living things is found in their agency and not their passivity. It lies in their ability to discriminate, move, sense, communicate, relate, adapt, coordinate, and choose. Their power, in short, is in their ability to act in a manner that is neither predetermined nor random. It is this potency, this agential quality, that makes living things so attractive to engineers.

To this extent, I argue, the very heart of biotechnology's operation is marked by a double bind. The double bind works like this: biotech seeks to take from the living the potency of life, but it tries to do so in a way that keeps that potency from getting in the way of the highly controlled—namely, computational—operations that engineers seek to enact. In other words, technologists want the objects of biotechnology to act like living things—to communicate, discern, respond, adapt, et cetera—while nonetheless remaining passive at the level of the basic mechanisms that need to be engineered to make these objects work in a predictable and controlled manner.

I suspect that this double bind between agency and passivity will encumber biotech's ability to deliver the salvational futures it promises. The double bind is, for that reason, central to the concerns of this book. Which is another way of saying that because biotechnology has become a vital factor in the ordering of twenty-first-century life, this double bind constitutes a matter of political, legal, ecological, and ethical, and not only technical, concern.

References

Aristotle. 1984. "Generation of Animals." *The Complete Works of Aristotle*. Princeton, NJ: Princeton University Press.

Balmer, Andrew, Kate Bulpin, and Susan Molyneux-Hodgson. 2016. *Synthetic Biology: A Sociology of Changing Practices*. New York: Palgrave Macmillan.

Canguilhem, Georges. 2000. *A Vital Rationalist: Selected Writings of Georges Canguilhem*. Edited by Francois Delaporte. Cambridge, MA: Zone Books.

———. 2009. *Knowledge of Life*. Translated by Stefanos Geroulanos and Daniela Ginsburg. Brooklyn, NY: Fordham University Press.

Kohn, Eduardo. 2015. "Anthropology of Ontologies." *Annual Review of Anthropology* 44: 311–327.

Jasanoff, Sheila, J. Benjamin Hurlbut, and Krishanu Saha. 2015. "CRISPR Democracy: Gene Editing and the Need for Inclusive Deliberation." *Issues in Science and Technology* 32: 25–32.

Peirce, Charles. 2009. "On a New List of Categories (1867)." In *The Essential Peirce*, vol. 1. Edited by Nathan Houser and Christian Kloesel. Bloomington, IN: Indiana University Press.

Riskin, Jessica. 2016. *The Restless Clock: A History of the Centuries-Long Argument over What Makes Living Things Tick*. Chicago, IL: University of Chicago Press.

Spinoza, Baruch. 2002. *The Ethics*. In *Spinoza: Complete Works*. Edited by Michael L. Morgan. Indianapolis, IN/Cambridge, MA: Hackett Publishing Company.

Taylor, Charles. 2002. "Modern Social Imaginaries." *Public Culture* 14 (1) (Winter): 91–124.

Weber, Max. 1958. *From Max Weber: Essays in Sociology*. Edited by Gerth, Hans H. and C. Wright Mills. Oxford: Oxford University Press.

———. 2015. "Science as a Vocation." In *Science, Reason, Modernity: Readings for an Anthropology of the Contemporary*. Edited by Anthony Stavrianakis, Gaymon Bennett, and Lyle Fearnley, 45–74. Brooklyn, NY: Fordham University Press.

Afterword

Governing Gene Editing
A Constitutional Conversation

Stephen Hilgartner

In explaining the decision to name CRISPR the "2015 Breakthrough of the Year," the journal *Science* declared: "For better or worse, we all now live in CRISPR's world" (Travis 2015). Such statements capture the sense of irreversible transformation that surrounds this new and powerful technology. How radical the changes stemming from gene editing will ultimately prove to be remains uncertain. But if we set aside this caveat and take claims of revolution seriously, a question immediately emerges: *What sort of new world have we entered?*

One answer: a world of dilemmas and questions. If humans have achieved unprecedented control over the basic material of life, how should they employ this power? If the human genome can now be edited, should we change the human germline? If genes can be driven into wild populations, should we use this capacity to edit the natural environment?

"CRISPR's world" is also a world of promises and warnings, and commentators who delineate the unprecedented opportunities and perilous threats generally have no trouble identifying issues for "society" to resolve. But what precisely is this "society," and through what mechanisms do we expect it to "resolve" such issues? Governance in this new world seems to have become strangely unmoored. Not only do central categories of social order such as the "human" and the "natural" seem to be losing some of their foundational status, but finding or constituting forums where politically legitimate settlements can be reached looks challenging. If gene drives introduce changes that, to use Irus Braverman's phrase, "spread like fire" across jurisdictions (Chapter 3, page 56, this volume), then what would be the appropriate site of decision making, or of establishing appropriate societal control over this emerging technology (Joly 2015)?

At the workshop where the chapters that appear in this volume were first presented, one audience member commented that she was "blown away," not only by the speed with which gene editing technologies are taking shape, but also by the invisibility of the ongoing debates about their use and regulation. She was troubled that decisions about whether and how we should edit nature were being made in institutions and ad hoc spaces largely outside of public view, and she wondered aloud how it could be that decisions of such gravity were being settled through informal and quasi-formal processes, without established mechanisms for broader and more meaningful public participation.

"CRISPR's world" is clearly a world in-the-making. What it will become cannot now be known, but this new world is being constituted through a process now unfolding in a variety of institutional and ad hoc settings. The chapters in this volume paint a fascinating picture of the scientific and social engineering that is underway. The diverse backgrounds of the authors—in science, law, philosophy, science and technology studies, and religious studies—and the quality of the chapters give the collection impressive breadth and depth. Each chapter is carefully argued and interesting, and each addresses an important topic in a sophisticated and often provocative way. Taken together, they offer a cutting-edge view of the debate about the governance of CRISPR, and it is an honor to comment on them in this Afterword. Rather than respond to each chapter individually, I will briefly explore what comparing modes of reasoning across the collection as a whole can tell us about the nature of the debate about governing CRISPR.

Visions of Governance

A central theme of this volume is the idea that science and technology need to be understood as a mode of government, an idea developed throughout the work of Sheila Jasanoff, a scholar in the field of Science and Technology Studies (STS). In an article reflecting on technology and politics at the turn of the millennium, Jasanoff (2003) argues that the contemporary world has entered what legal and political theorists might call a "constitutional moment"—a time in which the basic rules and infrastructures that establish order are being established or fundamentally changed. This description clearly applies to the governance of gene editing today. This world is very much in the process of being constituted, technologically (as techniques and applications proliferate), sociopolitically (as regimes for fitting gene editing into social orders begin to take shape), and biologically (as the prospects of gene editing to remake organisms and ecosystems are explored).

From such a perspective, the chapters of this volume can be read as what we might call a "constitutional conversation"; that is, an exchange in which the basic principles of governance are articulated and debated. Although mere conversations lack decision-making capacity and a formal legal standing, they are nevertheless spaces where frameworks of governance are configured and advanced. Moreover, in some cases, such conversations make strides toward creating the shared ontologies that underwrite the frameworks that, in turn, constitute new regimes of governance. Reading the collection in this way brings to the fore how each chapter differently frames problems of governance.

Alexander Travis's analysis of ethical issues in gene editing in dogs implicitly frames good governance as grounded in "right" reasoning: the problem of governance is construed as logically evaluating technological possibilities in order to draw lines separating permissible from impermissible technologies and applications (Chapter 8). Travis carefully evaluates the promise of gene editing in dogs, considering risks and such goals as improving canine health and gaining knowledge and experience that may someday be useful for human genome editing. The chapter also articulates principles, quoting the Veterinarian's Oath, and using it to define specific lines that should not be crossed, such as editing genomes in ways that would cause suffering or creating "imaginary animals," like dogs with zebra stripes. Throughout the chapter, Travis lays out key scientific facts and makes technical distinctions, some of which—such as the difference between gene editing and gene repair (a subset of the gene editing category)—become moral guideposts. This argument advances a particular vision of proper order, grounded in a hybrid of scientific and moral reasoning, through a process Jasanoff (2004) calls interactional coproduction. The overall picture that emerges is of a promising technology susceptible to guidance by a partnership of science and ethics. This mode of reasoning is central to bioethics and its efforts to govern emerging biotechnologies.

Ronald Sandler's chapter on the ethics of using gene drive as a species conservation tool explores questions about the acceptability of using gene drive to promote conservation goals, for example by modifying a wild species to adapt it to climate change (Chapter 2). Like Travis, Sandler's chapter frames governance problems mainly as a matter of "right" reasoning. He argues that categorical judgments about the acceptability of using gene editing in conservation cannot be justified at the level of the technique or strategy; instead, a case-by-case inquiry is needed about each concrete proposal. To aid in evaluating cases, Sandler offers a set of "criteria" to provide guidance. But Sandler also notes the existence of controversy, procedural issues, and concerns about "playing god" (Chapter 2, page 45). For Sandler, such matters as community concerns could thus become a potentially important dimension of governance.

Like Sandler and Travis, Todd Kuiken addresses the uses of gene drive in conservation, but his analysis frames governance less as an issue of "right" reason than as a question of material control over technological capacity. As an environmentalist, Kuiken is ambivalent about using gene drives as a conservation tool. He is also somewhat impatient with the failure of state authorities to institute meaningful environmental protections. In this context, the prospect that DIY biotechnology could enable environmental "vigilantes" to take matters into their own hands emerges as a distinct possibility. But marginalized environmental activists are not the only actors potentially interested in using gene drives to alter ecosystems. Kuiken describes research programs supported by the United States Department of Defense to explore the use of gene drives to protect crops or to enroll engineered insects as "allies" in biosecurity. One question that emerges from Kuiken's chapter is this: how feasible would it be to institute effective controls on editing the environment once this "genie is out of the bottle," so to speak?

In contrast to Travis and Sandler, Stuart Newman describes a world in which social forces, especially the logic of commercial opportunities in biotechnology, override moral imperatives (Chapter 7). Indeed, his chapter betrays skepticism about the efficacy of bioethical reasoning and regulatory institutions to maintain moral boundaries. Newman contends that bioethics has failed to govern biotechnology in the context of the human embryo. He believes that germline engineering of humans is a fundamental moral boundary that must not be crossed. Nonetheless, he forcefully argues that such moral lines are unable to withstand the pressure from the logic of contemporary research practices. As an example, he describes how he filed a patent application in 1997 on chimeric human-animal embryos and on animals containing human and animal cells. Newman's intent was not to establish ownership over the chimera, but to call public attention to the ethical problems inherent in such creations. This move, he writes, inspired law review articles and raised "significant consternation" in the scientific community (Chapter 7, page 142). Prominent scientists argued that no one would ever want to make such chimeras. Yet within a decade, one group had produced a mouse blastocyst containing human cells, and a second group had created a sheep with partly human organs. Such experiences suggest to Newman that, although restrictions on human germline modification are needed, as a practical matter, we will not be able to make them stick.

Newman's filing a patent application on human-animal chimeras can be regarded as a piece of conceptual art or as a performance, as Irus Braverman notes (Introduction, page 8). Along similar lines, Lori Andrews surveys recent literary and artistic works that use irony and satire, expose contradictions, problematize the normal, or suggest new, creative ways to deploy biotechnology in service of human needs. She argues that artistic expression can provide insightful provocations to inform public policy discussion, suggesting that artists can forcefully raise issues that may be more difficult to present in the rational and legal modes of argumentation that prevail in bioethics commissions and courts (Chapter 6).

In contrast to the authors discussed so far, Irus Braverman, J. Benjamin Hurlbut, and Gaymon Bennett (in Chapters 3, 4, and 9, respectively) approach CRISPR from a reflexive perspective that treats the forms of reasoning that are shaping the gene editing debate as their central subject. Each focuses on reasoning at a different scale. Braverman takes an ethnographic look at how individual scientists reason about gene editing. Her focus is on how several individuals (among them Kevin Esvelt, the author of Chapter 1) frame their own ethical stance vis-à-vis the natural, the human, and the nonhuman animal. She also considers affect. Her interviews shed considerable light on the scientific self and, in particular, on the relationship of the scientific self to questions of self-governance by scientists. In Esvelt's case, for example, she illuminates how his philosophical commitment to a strong utilitarianism, coupled with contempt for romanticized views of nature, have led him to perceive a moral imperative to reengineer nature for the purpose of relieving human and nonhuman suffering.

In his chapter, J. Benjamin Hurlbut also takes a reflexive stance. His focus, however, is not so much toward the individual scientist but rather toward a vital, if often unnoticed, part of the regulatory machinery that has governed biotechnology since the recombinant DNA debate of the 1970s (Chapter 4). Hurlbut analyzes what he calls "the imaginary of biological containment"—the idea of building controls directly into genetically modified organisms to discipline them and thus to prevent them from causing harm. This sociotechnical imaginary

(Jasanoff and Kim 2015; Hilgartner 2015), he argues, is a mode of governance that narrows conceptions of the risk of biotechnology, thereby maintaining an asymmetry that treats future benefits as tangible and future risks as speculative.

At the outset of his chapter, which wraps up the volume, Gaymon Bennett argues that biotechnology "needs to be governed at the level of the broad cultural imaginaries," not in a piecemeal manner (Chapter 9). Central to what he calls the "biotechnical imaginary" is the "disenchantment" of living things. Where other authors look to draw ethical lines, regulate risks, or establish procedural mechanisms to regulate gene editing, Bennett sees the failures of governance as stemming from a deeper cause in the conceptual apparatus with which we apprehend the world. He sees the possibility for such change in the realm of digital biology, which he believes has the potential to unsettle the regnant biotechnical imaginary through its struggle to disenchant the living world. Digital biology, he argues, "proceeds on the assumption that life can be made to operate as a set of passive mechanisms." In contrast, gene editing via CRISPR "approaches living things as self-regulating complexities" (Chapter 9, page 176). Yet these potential tensions between CRISPR and digital biology were resolved in a political manner: "the CRISPR breakthroughs were almost immediately taken up by the digital biology publicity machinery, and made to reinforce the already-diffuse sense that biological engineering has become a digital designer's playground" (ibid.).

Kevin Esvelt's chapter (Chapter 1) deserves a lengthy discussion in the context of my choice of theme for this Afterword. In this chapter, Esvelt lays out a particularly ambitious—and some might say hubristic—vision. Whereas a scientist like Travis suggests that gene editing can be governed by well-institutionalized modes of bioethical reasoning, Esvelt seeks to radically restructure how the scientific community operates. His chapter presents a call for implementing fundamental change in scientific institutions, and he describes the tangible work that he is pursuing to constitute what amounts to a new mode of governance.

Esvelt, who was prominently involved in developing gene drive, envisions using the technology to restructure social order both within and beyond the scientific community. His goals include ensuring biosafety, avoiding public relations disasters that would generate a backlash against biotechnology, and making scientific research more transparent by mandating new forms of openness. In part, he seeks transparency to enhance biosafety on the theory that when many eyes scrutinize proposed experiments, risks are more likely to be anticipated. He also fears that secrecy might provoke public distrust in science, and that if a gene drive project conceived in the privacy of the individual scientist's laboratory were to go awry, the public reaction would slow progress in biotechnology for many years to come. Esvelt finally seeks to build biosafety directly into gene drive systems using "daisy drives" that will limit their capacity to spread through a novel form of biological containment.

Yet Esvelt seems most excited by the prospect of using gene drive technology as a "catalyst" to reengineer the scientific community (Chapter 1, page 29). Here, his goals are social, not biological: he seeks to combat secrecy in the research system, eliminating a "prisoner's dilemma" that retards research and leads to duplication of effort (Chapter 3). His chapter presents an explicit argument for significant constitutional change intended to make scientific research more open, transparent, and rational, and he sketches the outlines of a regime that would bring openness to a scientific community accustomed to conducting its research behind closed doors.

Esvelt's proposed regime is intended to have jurisdiction over all gene drive experiments. The framework that constitutes its mode of governance centers on five main types of agents. Four of these agents—scientific journals, funders, policy makers, and a nonprofit that holds key intellectual property—would work to impose openness on the fifth: researchers, each of whom would be required to make proposals for gene drive research public to allow collective assessment of risk (Chapter 1, pages 28, 30). Briefly, the scheme calls for convincing funders to mandate public disclosure of grant proposals and persuading journals to require researchers to preregister gene drive projects as a condition of future publication. Esvelt also seeks to back the regime using the power of intellectual property claims on gene drives that he filed with Harvard and MIT. This would be accomplished by issuing licenses to use gene drive technology only to researchers who have agreed to openness throughout the research process. Moreover, he sees achieving openness in the gene drive domain as only a first step. It would provide a model that could spread to other areas of science, inspiring a radical reordering of how science is done.

Beyond regulating risks or realizing the benefits of biotechnology, Esvelt envisions what amounts to a constitutional change in the scientific community, with significant reallocation of authority and of the rights, duties, privileges, and powers of the organizations and researchers implicated in the regime. This rather ambitious proposal nicely illustrates the social engineering that scientists and others operating at the frontiers of emerging technologies often attempt. It also shows how these regimes typically rely on a mixture of discursive, technological, and legal means to establish control. Thus, Esvelt's proposed regime relies on an assemblage of legal mechanisms (e.g., patents and contracts), organizational policies (e.g., of journals and funders), technologies (e.g., daisy drives), and ontologies (e.g., shared definitions of "gene drive technology" and shared concepts of "acceptable risk"). It also draws support from widely shared sociotechnical imaginaries, such as the imaginary of scientific progress as a source of prosperity, health, and human flourishing and the imaginary of scientific openness. Not least, the regime relies on what J. Benjamin Hurlbut (Chapter 5, page 86) calls "biological containment as a mode of governance," and on the idea that the scientific community should properly settle technical questions.

Esvelt's proposed regime is only one of many visions for the project of governing gene editing. Since many more visions of radical change are proposed than actually take place, we might ask: how likely it is that Esvelt's vision will be successfully implemented? While precise predictions are impossible, the coproduction approach is able to provide some promising directions of inquiry. In particular, it suggests that visions that are compatible with the conceptual and institutional furnishings of existing social orders are more likely to survive and remain true to their original form. Radical changes, whether political or technological, do not always unfold in the ways that their leaders initially hoped they would. Given the extent to which structures that inhibit scientific openness are embedded in contemporary institutions and in the identities of scientists, efforts to implement Esvelt's vision will likely face an uphill battle, just as some of the more radical attempts to reorder science did during the Human Genome Project (Hilgartner 2017; see also Chapter 4). Moreover, if building gene drives turns out to be easy enough for widespread DIY use, it seems unlikely that a regime like that imagined by Esvelt will be capable of containing the enthusiasm of such actors as the environmental vigilantes that Kuiken describes and anticipates (Chapter 5).

Looking to the Future

Read as a constitutional conversation, this cutting-edge collection reveals substantial diversity in what governing gene editing means to specialists in the field today. The challenge of negotiating sustainable settlements given differences in normative goals will be substantial, and there are obviously no simple checklists of "best practices" to guide us through the choices ahead. The existence of such conversations is the only reason for hope that modes of governing gene editing that are wise, inclusive, and democratic will emerge.

References

Hilgartner, Stephen. 2015. "Capturing the Imaginary: Vanguards, Visions, and the Synthetic Biology Revolution." In *Science and Democracy: Making Knowledge and Making Power in the Biosciences and Beyond*. Edited by Stephen Hilgartner, Clark Miller, and Rob Hagendijk, 33–55. New York: Routledge.

———. 2017. *Reordering Life: Knowledge and Control in the Genomics Revolution*. Cambridge, MA: MIT Press.

Jasanoff, Sheila. 2003. "In a Constitutional Moment: Science and Social Order at the Millennium." In *Social Studies of Science and Technology: Looking Back, Ahead; Yearbook of the Sociology of the Sciences*. Edited by Bernward Joerges and Helga Nowotny, 155–180. Dordrecht, NL: Kluwer Academic.

———. 2004. "Ordering Knowledge, Ordering Society." In *States of Knowledge: The Co-Production of Science and Social Order*. Edited by Sheila Jasanoff, 13–45. New York: Routledge.

Jasanoff, Sheila and Sang-Hyun Kim, eds. 2015. *Dreamscapes of Modernity: Sociotechnical Imaginaries and the Fabrication of Power*. Chicago: University of Chicago Press.

Joly, Pierre-Benoît. 2015. "Governing Emerging Technologies? The Need to Think Outside the (Black) Box." In *Science & Democracy: Making Knowledge and Making Power in the Biosciences and Beyond*. Edited by Stephen Hilgartner, Clark Miller, and Rob Hagendijk, 133–155. New York: Routledge.

Travis, John. 2015. "Making the Cut." *Science* 350: 1456–1457.

Index